OPTIMAL RESOURCE ALLOCATION

OPTIMAL RESOURCE ALLOCATION

With Practical Statistical Applications and Theory

Igor A. Ushakov

WILEY

For general information on our other products and services or for technical support, please contact our Customer Care Department within the United States at (800) 762-2974, outside the United States at (317) 572-3993 or fax (317) 572-4002.

Wiley also publishes its books in a variety of electronic formats. Some content that appears in print may not be available in electronic formats. For more information about Wiley products, visit our web site at www.wiley.com.

Library of Congress Cataloging-in-Publication Data

Ushakov, I. A. (Igor Alekseevich)
    Optimal resource allocation : with practical statistical applications and theory / Igor A. Ushakov.
        pages   cm
    Includes bibliographical references and index.
    ISBN 978-1-118-38997-3 (cloth)
    1. Resource allocation–Statistical methods.   I. Title.
    T57.77.U84 2013
    658.4′033–dc23
                                2012040258

Printed in the United States of America

10 9 8 7 6 5 4 3 2 1

In memory of John D. Kettelle, Jr.,
my friend, colleague, and informal teacher

CONTENTS

PREFACE

Optimal resource allocation is an extremely important part of many human activities, including reliability engineering. One of the first problems that arose in this engineering area was optimal allocation of spare units. Then it came to optimization of networks of various natures (communication, transportation, energy transmission, etc.) and now it is an important part of counter-terrorism protection.

Actually, these questions have always stood and still stand: How can one achieve maximum gain with limited expenses? How can one fulfill requirements with minimum expenses?

In this book is an overview of different approaches of optimal resource allocation, from classical LaGrange methods to modern heuristic algorithms.

This book is not a tutorial in the common sense of the word. It is not a reliability "cookbook." It is more a bridge between reliability engineering and applied mathematics in the field of optimal allocation of resources for systems' reliability increase. It supplies the reader with basic knowledge in optimization theory and presents examples of application of the corresponding mathematical methods to real world problems. The book's objective is to inspire the reader to visit the wonderful area of applied methods of optimization, rather than to give him or her a mathematical course on optimization.

Examples with sometimes tedious and bulky numerical calculations should not frighten the reader. They are given with the sole purpose of demonstrating "a kitchen" of calculations. All these calculations have to be performed by a computer. Optimization programs themselves are simple enough. (For instance, all numerical examples were performed with the help a simple program in MS Office Excel.) At the very end of the book there is a complete enough list of monographs on the topic.

Who are potential readers of the book? First of all, engineers who design complex systems and mathematicians who are involved in "mathematical support" of engineering projects. Another wide category is college and university students, especially before they take classes on optimization theory. Last, university professors could use the material in the book, taking numerical examples and case studies for illustration of the methods they are teaching.

In conclusion, I would like to say a few words about references at the end of chapters. Each of them is not a list of references, but rather a bibliography presented in a chronological order. The author's belief is that such a list will allow the reader to trace the evolution of the considered topic. The lists, of course, are not full, for which the author apologizes in advance. However, as Kozma Prutkov (a pseudonym for a group of satirists at the end of the 19th century) said: "Nobody can embrace the unembraceable."

I would like to express my deep gratitude to my friend and colleague Dr. Gregory Levitin, who supplied me with materials on genetic algorithms and optimal redundancy in multi-state systems. I also would like to thank my friend Dr. Simon Teplitsky for scrupulously reading the draft of the book and giving a number of useful comments.

This book is in memory of my friend and colleague Dr. John D. Kettelle, a former mariner who fought in WWII and later made a significant input in dynamic programming. His name was known to me in the late 1960s when I was a young engineer in the former Soviet Union. I had been working at one of the R&D institutes of

the Soviet military–industrial establishment; my duty was project-
ing spare stocks for large scale military systems.

I met Dr. J. Kettelle in person in the early 1990s when I came
to the United States as Distinguished Visiting Professor at The
George Washington University. After two years at the university,
I was invited by John to work at Ketron, Inc., the company that he
established and led. We became friends.

I will remember John forever.

IGOR A. USHAKOV

San Diego, California

CHAPTER 1

BASIC MATHEMATICAL REDUNDANCY MODELS

A series system of independent subsystems is usually considered as a starting point for optimal redundancy problems. The most common case is when one considers a group of redundant units as a subsystem. The *reliability objective function* of a series system is usually expressed as a product of probabilities of successful operation of its subsystems. The *cost objective function* is usually assumed as a linear function of the number of system's units.

There are also more complex models (multi-purpose systems and multi-constraint problems) or more complex objective functions, such as average performance or the mean time to failure. However, we don't limit ourselves to pure reliability models. The reader will find a number of examples with various networks as well as examples of resource allocation in counter-terrorism protection.

In this book we consider main practical cases, describe various methods of solutions of optimal redundancy problems, and

Optimal Resource Allocation: With Practical Statistical Applications and Theory,
First Edition. Igor A. Ushakov.
© 2013 John Wiley & Sons, Inc. Published 2013 by John Wiley & Sons, Inc.

demonstrate solving the problems with numerical examples. Finally, several case studies are presented that reflect the author's personal experience and can demonstrate practical applications of methodology.

1.1 TYPES OF MODELS

A number of mathematical models of systems with redundancy have been developed during the roughly half a century of modern reliability theory. Some of these models are rather specific and some of them are even "extravagant." We limit ourselves in this discussion to the main types of redundancy and demonstrate on them how methods of optimal redundancy can be applied to solutions of the optimal resource allocation. Redundancy in general is a wide concept, however, we mainly will consider the use of a redundant unit to provide (or increase) system reliability.

Let us call a set of operating and redundant units of the same type *a redundant group*. Redundant units within a redundant group can be in one of two states: active (in the same regime as operating units, i.e., so-called hot redundancy) and standby (idle redundant units waiting to replace failed units, i.e. so-called cold redundancy). In both cases there are two possible situations: failed units could be repaired and returned to the redundant group or unit failures lead to exhaustion of the redundancy.

In accordance with such very rough classifications of redundancy methods, this chapter structure will be arranged as presented in Table 1.1.

We consider two main reliability indices: probability of failure-free operation during some required fixed time t_0, $R(t_0)$, and mean time to failure, T. In practice, we often deal with a system consisting of a serial connection of redundant groups (see Fig. 1.1). Usually, such kinds of structures are found in systems with spare stocks with periodical replenishment.

TABLE 1.1 Types of Redundancy

		1. Redundant units regime	
		Active	Standby
2. Type of maintenance	Non-repairable	Section 1.1	Section 1.2
	Repairable	Section 1.3	Section 1.4

FIGURE 1.1 General block diagram of series connection of redundant groups.

FIGURE 1.2 Block diagram of a duplicated system.

1.2 NON-REPAIRABLE REDUNDANT GROUP WITH ACTIVE REDUNDANT UNITS

Let us begin with a simplest redundant group of two units (duplication), as in Figure 1.2.

Such a system operates successfully if at least one unit is operating. If one denotes random time to failure of unit k by ξ_k, then the system time to failure, ξ, could be written as

$$\xi = \max\{\xi_1, \xi_2\}. \tag{1.1}$$

The time diagram in Figure 1.3 explains Equation (1.1).

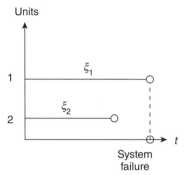

FIGURE 1.3 Time diagram for a non-repairable duplicated system with both units active.

The probability of failure-free operation (PFFO) during time t for this system is equal to

$$R(t) = 1 - [1 - r(t)]^2, \tag{1.2}$$

where $r(t)$ is PFFO of a single active unit.

We will assume an exponential distribution of time to failure for an active unit:

$$F(t) = \exp(-\lambda t). \tag{1.3}$$

In this case the mean time to failure (MTTF), T, is equal to:

$$T = E\{\xi\} = E\{\max(\xi_1, \xi_2)\} = \int_0^\infty R(t)dt$$

$$= \int_0^\infty 1 - [1 - \exp(\lambda t)]^2 dt = (1 + 0.5) \cdot \frac{1}{\lambda}. \tag{1.4}$$

Now consider a group of n redundant units that survives if at least one unit is operating (Fig. 1.4).

FIGURE 1.4 Block diagram of redundant group of *n* active units.

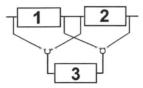

FIGURE 1.5 Block diagram of a "2 out of 3" structure with active redundant unit.

We omit further detailed explanations that could be found in any textbook on reliability (see Bibliography to Chapter 1).

For this case PFFO is equal:

$$R(t) = 1 - [1 - r(t)]^n, \tag{1.5}$$

and the mean time to failure (under assumption of the exponential failure distribution) is

$$T = \sum_{1 \leq k \leq n} \frac{1}{k}. \tag{1.6}$$

The most practical system of interest is the so-called *k* out of *n* structure. In this case, the system consists of *n* active units in total. The system is deemed to be operating successfully if *k* or more units have not failed (sometimes this type of redundancy is called "floating"). The simplest system frequently found in engineering practice is a "2 out of 3" structure (see Fig. 1.5).

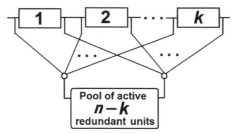

FIGURE 1.6   Block diagram of a "k out of n" structure with active redundant units.

A block diagram for general case can be presented in the following conditional way. It is assumed that any redundant unit can immediately operate instead of any of k "main" units in case a failure.

Redundancy of this type can be found in multi-channel systems, for instance, in base stations of various telecommunication networks: transmitter or receiver modules form a redundant group that includes operating units as well as a pool of active redundant units.

Such a system is operating until at least k of its units are operating (i.e., less than $n - k + 1$ failures have occurred). Thus, PFFO in this case is

$$R(t) = \sum_{k \leq j \leq n} \binom{n}{j} [p(t)]^{j} [1 - p(t)]^{n-j} \qquad (1.7)$$

and

$$T = \frac{1}{\lambda} \sum_{k \leq j \leq n} \frac{j}{n}. \qquad (1.8)$$

If a system is highly reliable, sometimes it is more reasonable to use Equation (1.7) in supplementary form (especially for approximate calculations when $p(t)$ is close to 1).

$$R(t) = 1 - \sum_{n-k+1 \le j \le n} \binom{n}{j} [1 - p(t)]^j [p(t)]^{n-j} \approx 1 - \binom{n}{n-k+1} [1 - p(t)]^{n-k+1}.$$

(1.9)

1.3 NON-REPAIRABLE REDUNDANT GROUP WITH STANDBY REDUNDANT UNITS

Again, begin with a duplicated system presented in Figure 1.7. For this type of system, the random time to failure is equal to:

$$\xi = \xi_1 + \xi_2.$$ (1.10)

The time diagram in Figure 1.8 explains Equation (1.10). The PFFO of a considered duplicate system can be written in the form:

$$R(t) = p_0(t) + p_1(t),$$ (1.11)

FIGURE 1.7 A non-repairable duplicated system with a standby redundant unit. (Here gray color denotes a standby unit.)

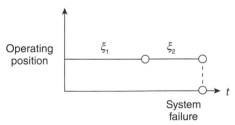

FIGURE 1.8 Time diagram for a non-repairable duplicated system with a standby redundant unit.

where $p_0(t)$ is the probability of no failures at time interval $[0, t]$, and $p_1(t)$ is the probability of exactly one failure in the same time interval. Under assumption of exponentiality of the time-to-failure distribution, one can write:

$$p_0 = \exp(-\lambda t) \tag{1.12}$$

and

$$p_1 = \lambda t \exp(-\lambda t), \tag{1.13}$$

so finally

$$R(t) = \exp(-\lambda t) \cdot (1 + \lambda t). \tag{1.14}$$

Mean time to failure is defined as

$$T = E\{\xi_1 + \xi_2\} = \frac{2}{\lambda}, \tag{1.15}$$

since $\lambda = 1/T$.

For a multiple standby redundancy, a block diagram can be presented in the form shown in Figure 1.9. For this redundant group, one can easily write (using the arguments given above):

$$R(t) = \exp(-\lambda t) \sum_{1 \le j \le n-1} \frac{(\lambda t)^j}{j!} \tag{1.16}$$

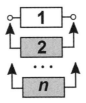

FIGURE 1.9 Block diagram of redundant group of one active and $n - 1$ standby units. (Here gray boxes indicate standby units.)

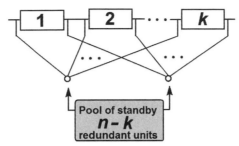

FIGURE 1.10 Block diagram of a "k out of n" structure with standby redundant units. (Here gray color is used to show standby redundant units.)

and

$$T = \frac{n}{\lambda}. \tag{1.17}$$

A block diagram for a general case of standby redundancy of k out of n type can be presented as shown in Figure 1.10. It is assumed that any failed operational unit can be instantaneously replaced by a spare unit. Of course, no replacement can be done instantaneously: in these cases, we keep in mind the five-second rule.[1]

This type of redundant group can be found in spare inventory with periodical restocking. Such replenishment is typical, for instance, for terrestrially distributed base stations of global satellite telecommunication systems. One observes a Poisson process of operating unit failures with parameter $k\lambda$, and the group operates until the number of failures exceeds $n - k$. The system PFFO during time t is equal to:

$$R(t) = \exp(-k\lambda t) \cdot \sum_{0 \le j \le n-k} \frac{(k\lambda t)^j}{j!} \tag{1.18}$$

[1]Russian joke: If a fallen object is picked up in 5 seconds, it is assumed that it hasn't fallen at all.

and the system MTTF is

$$T = \frac{1}{\lambda} \cdot \frac{n-k+1}{k}. \tag{1.19}$$

Of course, there are more complex structures that involve active and standby redundant units within the same redundant group. For instance, structure "k out of n" with active units could have additional "cold" redundancy that allows performing "painless" replacements of failed units.

1.4 REPAIRABLE REDUNDANT GROUP WITH ACTIVE REDUNDANT UNITS

Consider a group of two active redundant units, that is, two units in parallel. Each unit operates independently: after failure it is repaired during some time and then returns to its position. Behavior of each unit can be described as an alternating stochastic process: a unit changes its states: one of proper functionality during time ξ, followed by a failure state induced repair interval, η. The cycle of working/repairing repeats. This process is illustrated in Figure 1.11. From the figure, one can see that system failure occurs when failure intervals of both units overlap.

Notice that for repairable systems, one of the most significant reliability indices is the so-called availability coefficient, $\tilde{r}$. This reliability index is defined as the probability that the system is in a working state at some arbitrary moment of time. (This moment of time is assumed to be "far enough" from the moment the process starts.) It is clear that this probability for a single unit is equal to a portion of total time when a unit is in a working state, that is,

$$\tilde{r} = \frac{E\{\xi\}}{E\{\xi\} + E\{\eta\}}. \tag{1.20}$$

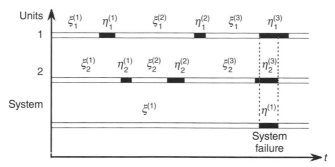

FIGURE 1.11 Time diagram for a repairable system with standby redundancy. White parts of a strip denote operating state of a unit and black parts its failure state. Here $\xi_j^{(i)}$ denotes jth operating interval of unit i, and $\eta_j^{(i)}$ denotes jth interval of repair of this unit.

If there are no restrictions, that is, each unit can be repaired independently, the system availability coefficient, $\tilde{R}$, can be written easily:

$$\tilde{R} = 1 - (1-r)^2. \tag{1.21}$$

For general types of distributions, reliability analysis is not simple. However, if one assumes exponential distributions for both ξ and η, reliability analysis can be performed with the help of Markov models.

If a redundant group consists of two units, there are two possible regimes of repair, depending on the number of repair facilities. If there is a single repair facility, units become dependent through the repair process: the failed unit can find the facility busy with the repair of a previously failed unit. Otherwise, units operate independently. Markov transition graphs for both cases are presented in Figure 1.12.

With the help of these transition graphs, one can easily write down a system of linear differential equations that can be used for obtaining various reliability indices. Take any two of the three equations:

FIGURE 1.12 Transition graphs for repairable duplicated system with active redundancy for two cases: restricted repair (only one failed unit can be repaired at a time) and unrestricted repair (each failed unit can be repaired independently). The digit in the circle denotes the number of failed units.

TABLE 1.2 Availability Coefficient for Two Repair Regimes

	Formula for availability coefficient, $\tilde{R}$	
	Restricted repair	Unrestricted repair
Strict formula	$\dfrac{1+2\gamma}{(1+\gamma)^2}$	$\dfrac{1+2\gamma}{(1+\gamma)^2+\gamma^2}$
Approximation for $\gamma \ll 1$	$1-\gamma^2$	$1-2\gamma^2$

$$\begin{cases} \dfrac{d}{dt}P_0(t) = -2\lambda P_0(t) + \mu P_1(t) \\[2mm] \dfrac{d}{dt}P_1(t) = 2\lambda P_0(t) - (\lambda + \mu)P_1(t) + \mu P_2(t) \\[2mm] \dfrac{d}{dt}P_2(t) = \lambda P_1(t) - \mu P_2(t) \text{ for restricted repair} \\[2mm] \text{or} \\[2mm] \dfrac{d}{dt}P_2(t) = \lambda P_1(t) - 2\mu P_2(t) \text{ for restricted repair} \end{cases} \qquad (1.22)$$

and take into account chosen initial conditions.

The availability coefficient for these two cases can be calculated using the formulas (where $\gamma = \lambda/\mu$) in Table 1.2. However, our intent is to present methods of optimal redundancy rather than to give detailed analysis of redundant systems. (Such analysis can be found almost in any book listed in the Bibliography to Chapter 1.) Thus

TABLE 1.3 Approximate Formulas for Availability Coefficient

Type of redundant group	Approximate formula for availability coefficient, $\tilde{R}$	
	Restricted repair	Unrestricted repair
Group of n units	$1 - (n!) \cdot \gamma^n$	$1 - \gamma^n$
Group of type "k out of n"	$1 - [(n-k+1)!] \cdot \binom{n}{n-k+1} \gamma^{n+1}$	$1 - \binom{n}{n-k+1} \gamma^{n+1}$

we will consider only the simplest models of redundant systems, that is, systems with unrestricted repair.

We avoid strict formulas because they are extremely clumsy; instead we present only approximate ones that mostly are used in practical engineering calculations (Table 1.3).

1.5 REPAIRABLE REDUNDANT GROUP WITH STANDBY REDUNDANT UNITS

Consider now a repairable group of two units: one active and one standby. Behavior of such a redundant group can be described with the help of a renewal process: after a failure of the operating unit a standby unit becomes the newly operating one, while the failed unit after repair becomes a standby one, and so on. System failure occurs when a unit undergoing repair is not ready to replace a now not operating unit that has just failed. The process of functioning in this type of duplicated system is illustrated in Figure 1.13. In this case, finding PFFO of the duplicated system is also possible with the use of Markov models under assumption of exponentiality of both distributions (of repair time and time to failure).

Transition graphs for restricted and unrestricted repair are shown in Figure 1.14.

Again, we present only approximate formulas in Table 1.4.

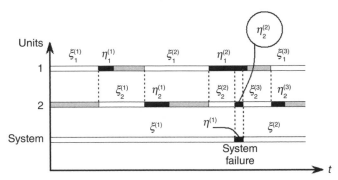

FIGURE 1.13 Time diagram for a repairable duplicated system with standby redundancy. White parts of a strip denote the operating state of a unit, gray parts show the standby state, and black parts show the failure state. Here $\xi_j^{(i)}$ denotes jth operating interval of unit i, and $\eta_j^{(i)}$ denotes jth interval of repair of this unit.

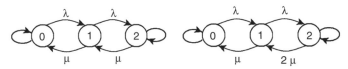

FIGURE 1.14 Transition graphs for repairable duplicated systems with standby redundancy for two cases: restricted repair (only one failed unit can be repaired at a time) and unrestricted repair (each failed unit can be repaired independently).

TABLE 1.4   Approximate Formulas for Availability Coefficient

Type of redundant group	Approximate formula for availability coefficient, $\tilde{R}$	
	Restricted repair	Unrestricted repair
Group of n units	$1 - \gamma^{n-k+1}$	$1 - \dfrac{\gamma^{n-k+1}}{(n-k+1)!}$
Group of type "k out of n"	$1 - (k\gamma)^{n-k+1}$	$1 - \dfrac{(k\gamma)^{n-k+1}}{(n-k+1)!}$

1.6 MULTI-LEVEL SYSTEMS AND SYSTEM PERFORMANCE ESTIMATION

Operation of a complex multi-level system cannot be satisfactorily described in traditional reliability terms. In this case, one has to talk about performance level of such systems rather than simple binary type "up and down" operating.

Let a system consist of n independent units characterized by their reliability indices $p_1, p_2, \ldots, p_n$. Assume that with unit failure a level of system performance degrades. Denote by Φ_i a quantitative measure of the system performance under the condition that unit i failed, by Φ_{ij} the same measure if units i and j failed, and in general, if some set of units, α have failed then the system performance is characterized by value Φ_α. In this case the system performance can be characterized by the mean value:

$$\Phi_{System} = \sum_{\alpha \in A} H_\alpha \Phi_\alpha, \qquad (1.23)$$

where A is a set of all possible states of units 1, 2, . . ., n, that is, power of this set is 2^n and

$$H_\alpha = \prod_{i \in \alpha} (1 - p_i) \prod_{i \in A \setminus \alpha} p_i, \qquad (1.24)$$

where notation $A \setminus \alpha$ means the total set of unit subscripts with exclusion of subset α.

The measure of system performance could be taken from conditional probability of successful fulfillment of the operation, productivity, or other operational parameters.

Several years after Kozlov and Ushakov (1966) had been published, there was a relative silence with quite rare appearance of works on the topic. Since average measure is not always a good characterization, soon there was a suggestion to evaluate the probability that multi-state system performance is exceeding some required level. In a sense, it was nothing more than introducing a failure criterion for a multi-state system. In this case, new formulation of the system reliability has the form

$$R_{System} = \Pr\{\Phi_\alpha \geq \Phi_{Required}\} = \sum_{\alpha:\Phi_\alpha \geq \Phi_{Required}} H_\alpha \Phi_\alpha. \qquad (1.25)$$

In 1985, Kurt Reinschke (in Ushakov, 1985) introduced a system that itself consists of multi-state units. However, this work also did not find an appropriate response among reliability specialists at the time.

Nevertheless, reliability analysis of multi-state systems has started for all three possible classes:

(1) Multi-state systems consisting of binary units

(2) Binary systems consisting of multi-state units

(3) Multi-state systems consisting of multi-state units.

In the late 1990s, there was a veritable avalanche of papers on this topic, which has maintained a steady flow ever since. This subject is considered in more detail in Chapter 11.

Naturally, after multi-system analysis, attention to the problems of optimal redundancy in such systems arose. Now the problem of optimal redundancy in multi-state systems is a subject of intensive research.

1.7   BRIEF REVIEW OF OTHER TYPES OF REDUNDANCY

In reliability theory, redundancy is understood as using additional units for replacement/substitution of failed units. Actually, there are many various types of redundancy. Below we briefly consider structural redundancy, functional redundancy, a system with spare time for operation performance, and so on.

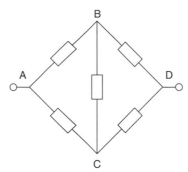

FIGURE 1.15   Bridge structure.

1.7.1   Two-Pole Structures

One of the typical types of structural redundancy is presented by networks. The simplest network structure is the so-called *bridge structure* (see Fig. 1.15). Assume that a connection between points A and D is needed.

A failure of any one unit does not lead to failure of the system because of the redundant structure. There are the following paths from A to D: *ABD, ACD, ACBD,* and *ABCD.* If at least one of those paths exists, the system performs its task. Of course, one can consider all cuts that lead to the system failure: *AB&AC, BD&CD, AB&BC&CD,* and *AC&BC&BD.* However, in this case we cannot use simple formulas of series and parallel systems, since paths are interdependent, as are cuts. Because of this, one can only write the upper and lower bounds for PFFO of such systems:

$$(1-Q_{AB}Q_{AC})\cdot(1-Q_{BD}Q_{CD})\cdot(1-Q_{AB}Q_{BC}Q_{CD})\cdot(1-Q_{AC}Q_{BC}Q_{BD}) < R_{Bridge}$$
$$< 1-(1-P_{AB}P_{BD})\cdot(1-P_{AC}P_{CD})\cdot(1-P_{AB}P_{BC}P_{CD})\cdot(1-P_{AC}P_{BC}P_{BD}).$$

$$(1.26)$$

For this simple case, one can find a strict solution using a straightforward enumeration of all possible system states:

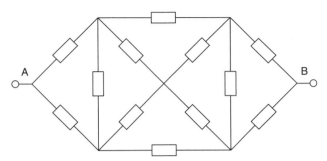

FIGURE 1.16   An example of a two-pole network.

$$R_{Bridge} = R_{BC}(1 - Q_{AB}Q_{AC}) \cdot (1 - Q_{BD}Q_{CD}) + Q_{BC}[1 - (1 - P_{AB}P_{BD})(1 - P_{AC}P_{CD})]. \tag{1.27}$$

More complex systems of this type are presented by the two-pole networks: in such systems a "signal" has to be delivered from terminal A to terminal B (see Fig. 1.16). Reliability analysis of such systems is normally performed with the use of Monte Carlo simulation.

For networks with a general structure, the exact value of the reliability index can be found only with the help of a direct enumeration. For evaluation of this index, one can use the upper and lower bounds of two types: Esary-Proschan boundaries (Barlow and Proschan, 1965) or Litvak-Ushakov boundaries (Ushakov, ed., 1985). Unfortunately, boundaries cannot be effectively used for solving optimal redundancy problems.

1.7.2   Multi-Pole Networks

This kind of network is very common in modern life, appearing in telecommunication networks, transportation and energy grids, and so on. The most important specific of such systems is their structural redundancy and the redundant capacity of their compo-

TABLE 1.5 Traffic in the Network (in conditional units)

	A	B	C	D
A	–	1	1	1
B	1	–	2	1
C	1	2	–	1
D	1	1	1	–

nents. We demonstrate the specifics of such systems using a simple illustrative example. Consider the bridge structure that was described above, but assume that each node is either a "sender" or a "receiver" of "flows" to each other. Of course, flows can be different, as well as the capacities of particular links. Assume that traffic is symmetrical, that is, traffic from X to Y is equal to traffic from Y to X. This assumption allows us to consider only one-way flow between any points.

Let the traffic in the considered network be described as is shown in Table 1.5. For normal operating, it is enough to have the capacities of the links as described in Figure 1.17. (We will assume that traffic within the network is distributed as uniformly as possible.)

However, links (as well as nodes) are subject to failure. For protection of the system against link failures, let us consider possible scenarios of link failure and measures of system protection by means of links' capacities increase.

What should we do if link AB has failed? The flow from A to B and from A to D should be redirected. Thus, successful operation of the network requires an increase of the links' capacities (see Fig. 1.18).

Since all four outside links are similar, failure of any link (AC, BD, or CD) leads to a similar situation. Thus, to protect the system against failure of any outside link, one should increase the capacities of each outside link from 2 to 3 units.

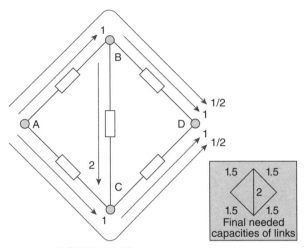

FIGURE 1.17 Traffic distribution.

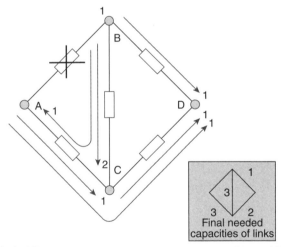

FIGURE 1.18 Traffic distribution in the case of link *AB* failure.

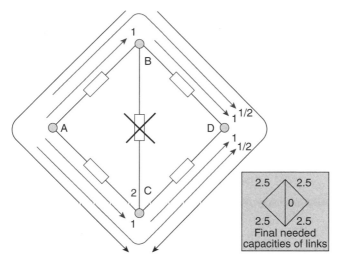

FIGURE 1.19   Traffic distribution in the case of link *BC* failure.

What happens if link *BC* fails? This link originally was used only for connecting nodes *B* and *C*. This traffic should be redistributed: half of the flow is directed through links *BA–AC*, and the rest through links *BD–DC*. To protect the system against link BC failure, the capacity of each outside link has to be increased by one unit.

To protect the system against any single link failure, one has to make link capacities corresponding to the maximum at each considered scenario, as demonstrated in Figure 1.20.

1.7.3   Branching Structures

Another rather specific type of redundant system is a branching structure system (see Fig. 1.21). In such systems, actual operational units are on the lowest level, and successfully operate only under the condition that their controlling units at the upper levels are successfully operating. Such structures are very common, especially in military control systems.

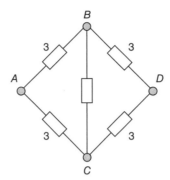

FIGURE 1.20 Final values of link capacities for a network protected against any possible single failure.

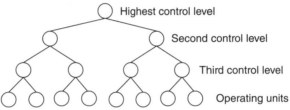

FIGURE 1.21 System with branching structure.

Assume that the branching system performs satisfactorily until four or more units of the lower level failed or lost control by upper level units. Types of possible system failures are given in Figure 1.22.

Of course, for complex systems the concept of "failure" is not adequate; instead, there is the notion of diminished performance. For instance, for the same branching system considered above, it is possible to introduce several levels of performance. Assume that the system performance depending on the system state is described by Table 1.6.

Usually, for such systems with structural redundancy, one uses the average level of performance. However, it is possible to introduce

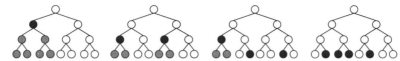

FIGURE 1.22 Types of situations when the branching system has 4 lower level units that have failed to perform needed operations. (Failed units are in black and units without control are in gray.)

TABLE 1.6 Levels of System Performance for Various System States

Quantity of failed units of lower level	Conditional level of performance
0	100%
1	99%
2	95%
3	80%
4	60%
5	50%
6	10%
7	2%
8	0%

a new failure criterion and talk about the reliability of such a system. For instance, under the assumption that admissible level of performance is 80%, one comes to the situation considered above: the system is considered failed only when four (or more) of its lower level units do not operate sufficiently (failed or lost control).

1.7.4 Functional Redundancy

Sometimes to increase the probability of successful performance of a system, designers envisage functional redundancy, that is, make it possible to use several different ways of completing a mission.

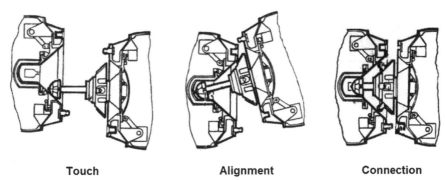

Touch   **Alignment**   **Connection**

FIGURE 1.23   Phases of a space shuttle docking to a space station.

As an example, one can consider the procedure of docking a space shuttle with a space station (Fig. 1.23).

This complex procedure can be fulfilled with the use of several various methods: by signals from the ground Mission Control Center (MCC), by the on-board computer system, and manually. In all these cases, video images sent from space objects are usually used. However, MCC can also use telemetry data. All methods can ensure success of the operation, though with different performance.

1.8   TIME REDUNDANCY

One very specific type of redundancy is the so-called *time redundancy*. There are three main schemes of time redundancy.

    (a) A system is operating during interval t_0. There are instantaneous interruptions of the system operation (failures), after which the system starts its operation from the beginning. The system operation is considered successful if during interval t_0 there is at least one interval with length larger than some required value τ. In other words, there is some extra time to restart the operation (see Fig. 1.24).

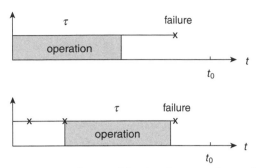

FIGURE 1.24 Examples of possible implementation of the successful system operation.

Denote the probability of success for such a system by $R(t_0 \mid \tau)$. If there is a failure on interval $[0, t_0]$ at such moment $x < \tau$ that still $t_0 - x > \tau$, the needed operation can be restarted, otherwise $R(t_0 \mid \tau) = 0$. This verbal explanation leads us to the recurrent expression

$$R(t_0 \mid \tau) = R(\tau) + \int_0^\tau R(x \mid t_0 - x) dF(x), \qquad (1.28)$$

where $F(x)$ is distribution function of the system time to failure.

These types of recurrent equations are usually solved numerically.

(b) Independent of the number of sustained failures, system operation is considered successful if the cumulative time of the system operation is no less than the required amount θ (see Fig. 1.25).

Denote the distribution of repair time, η, by $G(t)$. If the first failure has occurred at moment x such that $x > \theta$, it means that the system fulfilled its operation. If failure happens at moment ξ, the system can continue its operation after repair that takes time η, only if $t_0 - \eta > \theta$. It is clear

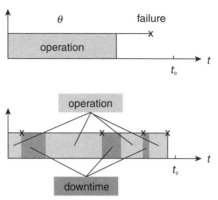

FIGURE 1.25 Examples of possible implementation of the successful system operation.

that the probability that the total operating time during interval $[0, t_0]$ is no less than θ is equal to the probability that the total repair time during the same interval is no larger than $t_0 - \theta$.

For this probability, one considers two events that lead to success:

- System works without failures during time θ from the beginning.
- System has failed at the moment $x < t_0 - \theta$, and was repaired during time y, and during the remaining interval of $t_0 - x - y$ accumulates $\theta - x$ units of time of successful operation. This verbal description permits us to write the following recurrent expression:

$$R(t_0|\theta) = 1 - F(t_0) + \int_0^{t_0}\left[\int_0^{t_0-x} R(t_0 - x - y|\theta)dG(y)\right]dF(x), \quad (1.29)$$

where $R(t_0 \mid z) = 0$ if $z < \theta$.

(c) A system "does not feel" failures of duration less than χ (Fig. 1.26). (In a sense, the system possesses a kind of "inertia" much like the famous "five second rule.")

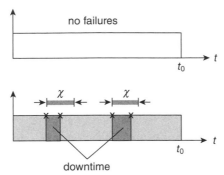

FIGURE 1.26 Time diagram for a system accumulating operation time.

A system is considered to be successfully operating if during period $[0, t_0]$ there is no down time longer than ψ. This case, in some sense, is a "mirror" of what was considered at the beginning. We will skip explanation details and immediately write the recurrent expression:

$$R(t_0 \mid \psi) = 1 - F(t_0) + \int_0^{t_0} \left[\int_0^{\eta} R(t_0 - x - y \mid \psi) dG(y) \right] dF(x). \quad (1.30)$$

We will not consider this type of redundancy in details; instead we refer the reader to special literature on the subject (Cherkesov, 1974; Kredentser, 1978).

1.9 SOME ADDITIONAL OPTIMIZATION PROBLEMS

1.9.1 Dynamic Redundancy

Dynamic redundancy models occupy an intermediate place between optimal redundancy and inventory control models.

The essence of a dynamic redundancy problem is contained in the following. Consider a system with n redundant units. Some redundant units are operating and represent an active redundancy. These units can be instantly switched into a working position without delay and, consequently, do not interrupt the normal operation of the system. These units have the same reliability parameters (e.g., for exponential distribution, and the same failure rate). The remaining units are on standby and cannot fail while waiting. But at the same time, these units can be switched in an active redundant regime only at some predetermined moments of time. The total number of such switching moments is usually restricted because of different technical and/or economical reasons.

A system failure occurs when at some moment there are no active redundant units to replace the main ones that have failed. At the same time, there may be many standby units that cannot be used because they cannot be instantly switched after a system failure.

Such situations in practice can arise in different space vehicles that are participating in long journeys through the Solar System. A similar situation occurs when one considers using uncontrolled remote technical objects whose monitoring and service can be performed only rarely.

It is clear that if all redundant units are switched to an active working position at an initial moment $t = 0$, the expenditure of these units is highest. Indeed, many units might fail in vain during the initial period. At the same time, the probability of the unit's failure during this interval will be small. On the other hand, if there are few active redundant units operating in the interval between two neighboring switching points, the probability of the system's failure decreases. In other words, from a general viewpoint, there should exist an optimal rule (program) of switching standby units into an active regime and allocating these units over all these periods.

Before we begin to formulate the mathematical problem, we discuss some important features of this problem in general.

Goal Function

Two main reliability indices are usually analyzed: the probability of failure-free system operation during some specified interval of time, and the mean time to system failure.

System Structure

Usually, for this type of problem, a parallel system is under analytical consideration. Even a simple series system requires a very complex analysis.

Using Active Redundant Units

One possibility is that actively redundant units might be used only during one period after being switched into the system. Afterward, they are no longer used, even if they have not failed. In other words, all units are divided in advance into several independent groups, and each group is working during its own specified period of time. After this period has ended, another group is switched into the active regime. In some sense, this regime is similar to the preventive maintenance regime.

Another possibility is to keep operationally redundant units in use for the next stages of operation. This is more effective but may entail some technical difficulties.

Controlled Parameters

As we mentioned above, there are two main parameters under our control: the moments of switching (i.e., the periods of work) and the number of units switched at each switching moment. Three particular problems arise: we need to choose the switching moments if the numbers of switched units are fixed in each stage; we need to choose the numbers of units switched in each stage if the switching moments are specified in advance; and, in general, we need to choose both the switching moments and the numbers of units switched at each stage.

Classes of Control

Consider two main classes of switching control. The first one is the so-called *prior rule* (*program switching*) where all decisions are made in advance at time $t = 0$. The second class is the *dynamic rule* where a decision about switching is made on the basis of current information about a system's state (number of forthcoming stages, number of standby units, number of operationally active units at the moment, etc.).

We note that analytical solutions are possible only for exponentially distributed TTFs. The only possible method of analysis for an arbitrary distribution is via a Monte Carlo simulation.

CHRONOLOGICAL BIBLIOGRAPHY OF MAIN MONOGRAPHS ON RELIABILITY THEORY (WITH TOPICS ON OPTIMIZATION)

Lloyd, D.K., and Lipov, M. 1962. *Reliaility Management, Methods and Mathematics.* Prentice Hall.

Barlow, R.E., and F. Proschan. 1965. *Mathematical Theory of Reliability.* John Wiley & Sons.

Kozlov, B.A., and Ushakov, I.A. 1966. *Brief Handbook on Reliability of Electronic Devices.* Sovetskoe Radio.

Raikin, A.L. 1967. *Elements of Reliability Theory for Engineering Design.* Sovetskoe Radio.

Polovko, A.M. 1968. *Fundamentals of Reliability Theory.* Academic Press.

Gnedenko, B.V., Belyaev, Y.K., and Solovyev, A.D. 1969. *Mathematical Methods in Reliability Theory.* Academic Press.

Ushakov, I.A. 1969. *Method of Solving Optimal Redundancy Problems under Constraints* (in Russian). Sovetskoe Radio.

Kozlov, B.A., and Ushakov, I.A. 1970. *Reliability Handbook.* Holt, Rinehart & Winston.

Cherkesov, G.N. 1974. *Reliability of Technical Systems with Time Redundancy* (in Russian). Sovetskoe Radio.

Barlow, R.E., and Proschan, F. 1975. *Statistical Theory of Reliability and Life Testing.* Holt, Rinehart & Winston.

Gadasin, V.A., and Ushakov, I.A. 1975. *Reliability of Complex Information and Control Systems* (in Russian). Sovetskoe Radio.

Kozlov, B.A., and Ushakov, I.A. 1975. *Handbook of Reliability Calculations for Electronic and Automatic Equipment* (in Russian). Sovetskoe Radio.

Kozlow, B.A., and Uschakow, I.A. 1978. *Handbuch zur Berehnung der Zuverlassigkeit in Elektronik und Automatechnik* (in German). Academie-Verlag.

Kredentser, B.P. 1978. *Forcasting Reliability for Time Redundamcy* (in Russian). Naukova Dumka.

Raikin, A.L. 1978. *Reliability Theory of Complex Systems* (in Russian). Sovietskoe Radio.

Kozlow, B.A., and Uschakow, I.A. 1979. *Handbuch zur Berehnung der Zuverlassigkeit in Elektronik und Automatechnik* (in German). Springer-Verlag.

Tillman, F.A., Hwang, C.L., and Kuo, W. 1980. *Optimization of System Reliability.* Marcel Dekker.

Barlow, R.E., and Proschan, F. 1981. *Statistical Theory of Reliability and Life Testing*, 2nd ed.

Gnedenko, B.V., ed. 1983. *Aspects of Mathematical Theory of Reliability* (in Russian). Radio i Svyaz.

Ushakov, I.A. 1983. *Textbook on Reliability Engineering* (in Bulgarian). VMEI.

Ushakov, I.A., ed. 1985. *Handbook on Reliability* (in Russian). Radio i Svyaz.

Rudenko, Y.N., and Ushakov, I.A. 1986. *Reliability of Power Systems* (in Russian). Nauka.

Reinschke, K., and Ushakov, I.A. 1987. *Application of Graph Theory for Reliability Analysis* (in German). Verlag Technik.

Reinschke, K., and Ushakov, I.A. 1988. *Application of Graph Theory for Reliability Analysis* (in Russian). Radio i Svyaz.

Reinschke, K., and Ushakov, I.A. 1988. *Application of Graph Theory for Reliability Analysis* (in German). Springer, Munchen-Vien.

Rudenko, Y.N., and Ushakov, I.A. 1989. *Reliability of Power Systems*, 2nd ed. (in Russian). Nauka.

Kececioglu, D. 1991. *Reliability Engineering Handbook.* Prentice-Hall.

Volkovich, V.L., Voloshin, A.F., Ushakov, I.A., and Zaslavsky, V.A. 1992. *Models and Methods of Optimization of Complex Systems Reliability* (in Russian). Naukova Dumka.

Ushakov, I.A. 1994. *Handbook of Reliability Engineering.* John Wiley & Sons.

Gnedenko, B.V., and Ushakov, I.A. 1995. *Probabilistic Reliability Engineering.* John Wiley & Sons.

Kapur, K.C., and Lamberson, L.R. 1997. *Reliability in Engineering Design.* John Wiley & Sons.

Gnedenko, B.V., Pavlov, I.V., and Ushakov, I.A. 1999. *Statistical Reliability Engineering.* John Wiley & Sons.

Kuo, W., and Zuo, M.J. 2003. *Optimal Reliability Modeling: Principles and Applications.* John Wiley & Sons.

Pham, H. 2003. *Handbook of Reliability Engineering.* Springer.

Kuo, W., Prasad, V.R., Tillman, F.A., and Hwang, C.-L. 2006. *Optimal Reliability Design: Fundamentals and Applications*. Cambridge University Press.

Levitin, G., ed. 2006. *Computational Intelligence in Reliability Engineering. Evolutionary Techniques in Reliability Analysis and Optimization*. Series: Studies in Computational Intelligence, vol. 39. Springer-Verlag.

Ushakov, I.A. 2007. *Course on Reliability Theory* (in Russian). Drofa.

Gertsbakh, I., and Shpungin, Y. 2010. *Models of Network Reliability*. CRC Press.

CHAPTER 2

FORMULATION OF OPTIMAL REDUNDANCY PROBLEMS

2.1 PROBLEM DESCRIPTION

One of the most frequently used methods of reliability increase is the use of additional (redundant) units, circuits, and blocks. This method is especially convenient when the principal solution of the system design has already been found: the use of redundant units usually does not cause a change in the overall structure of the system, but the use of extra units entails additional expense. Naturally, a system designer always tries to find the least expensive way to improve reliability. Thus, a designer faces two problems:

(1) *Direct problem of optimal redundancy*: Find such allocation of redundant units among different subsystems that warrants the required level of reliability index while using the smallest amount of resources possible.

Optimal Resource Allocation: With Practical Statistical Applications and Theory, First Edition. Igor A. Ushakov.
© 2013 John Wiley & Sons, Inc. Published 2013 by John Wiley & Sons, Inc.

(2) *Inverse problem of optimal redundancy*: Find such allocation of redundant units among different subsystems that maximizes the level of chosen reliability index under some specified constraints on the total cost of the system.

The choice of constraints depends on the specific engineering problem. Of course, the cost of a set of redundant units is not a unique objective function. For instance, for submarines the most serious constraint is the total volume (or weight) of spare units.

Consider a series system composed of n independent redundant groups (or subsystems). A redundant group is not necessarily a separate part of a system. In this context, this may be a group of units of the same type that use the same redundant units (see Fig. 2.1). For instance, in spare parts allocation problems, a redundant group might be a set of identical units located throughout the entire system in quite different places.

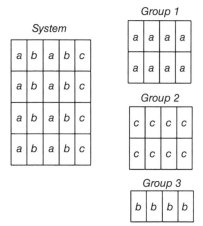

FIGURE 2.1 Modular system and "informal" redundant groups for this system.

2.2 FORMULATION OF THE OPTIMAL REDUNDANCY PROBLEM WITH A SINGLE RESTRICTION

The simplest (and, practically, the most often encountered) optimal redundancy problem is optimization of an objective function under a single constraint. Usually, the following two objective functions are considered: the cost of the total set of redundant units, $C(X)$, and the probability of a failure-free system operation, $R(X)$, where as above $X = (x_1, x_2 \ldots x_n)$ and x_i is the number of units within redundant group i.

The *direct problem of optimal redundancy* can be written as:

$$\min_{X}[C(X)\,|\,R(X) \geq R_0], \tag{2.1}$$

and the *inverse problem* can be written as:

$$\max_{X}[R(X)\,|\,C(X) \leq C_0], \tag{2.2}$$

where R_0 and C_0 are given constraints for the specific problems.

Cost of the redundant group as a whole is usually assumed as a linear function of number of redundant units and expressed as:

$$C(X) = C(x_1, x_2, \ldots, x_n) = \sum_{1 \leq i \leq n} C_i(x_i). \tag{2.3}$$

In most cases, we deal with a system that can be presented as a series connection of independent redundant groups. For such systems, the probability of a successful operation during time t, $R(t\,|\,X)$, and availability coefficient, $\tilde{R}(X)$, can be presented as a product of corresponding indices of redundant groups. Because both these objective functions are similar by their probabilistic nature, let us use a common notation $R(X)$ for both cases. Then we can write:

$$R(X) = R(x_1, x_2, \ldots, x_n) = \prod_{1 \leq i \leq n} R_i(x_i), \tag{2.4}$$

where $X = (x_1, x_2, \ldots, x_n)$ is the set of the system redundant units x_is of the ith type, $1 \leq i \leq n$, and the $R_i(x_i)$ are the reliability indices of the ith redundant group.

Sometimes it is more convenient to present Equation (2.3) in an additive form

$$L(X) = L(x_1, \ldots, x_n) = \sum_{1 \leq i \leq n} L_i(x_i) \tag{2.5}$$

where $L(X) = \ln R(X)$ and $L_i(x_i) = \ln L_i(x_i)$.

If a system is highly reliable, that is,

$$Q_i(x_i) = 1 - R_i(x_i) << \frac{1}{n} \tag{2.6}$$

or, equivalently,

$$Q(X) = Q(x_1, \ldots, x_n) << 1, \tag{2.7}$$

one can use the approximation

$$Q(X) = Q(x_i, \ldots, x_n) \approx \sum_{1 \leq i \leq n} Q_i(x_i). \tag{2.8}$$

Of course, similar problems can be formulated for other objective functions, for instance, for mean time to failure (or mean time between failures), T. Unfortunately, the calculation of $T(X)$ is usually rather difficult.

Example 2.1

Consider a simplest series system consisting of two units. Unit parameters are given in Table 2.1. We need to find:

(a) A number of units of both types, $X^{opt} = (x_1^{opt}, x_2^{opt})$ that satisfy the required level of system reliability index equal to 0.8 and deliver minimum possible system cost, and

TABLE 2.1   Unit Parameters for Example 2.1

	PFFO	Cost
Unit 1	0.7	1.2
Unit 2	0.6	2.7

TABLE 2.2   Values of System Cost for Various X

		x_2			
			1	2	3
			2.7	5.4	8.1
	1	1.2	3.9	6.6	9.3
	2	2.4	5.1	7.8	10.5
	3	3.6	6.3	9.0	11.7
	4	4.8	7.5	10.2	12.9
x_1	5	6.0	8.7	11.4	14.1

(b) A number of units of both types, x_1^{opt} and x_2^{opt}, that maximize system reliability index under constrain that the total system cost is not higher than 7 units.

It is assumed that "hot" redundancy is used for reliability improvement, that is, $R_i(x_i) = 1 - q_i^{x_i}$.

Since we don't assume any a priori knowledge of optimization methods, let us use a trivial enumerating method. For further convenience, let us introduce *triplets* that contain the following information: $\Delta_i(x_i) = \{x_i, R_i(x_i), C_i(x_i)\}$. We compile two tables with the system cost and probability of failure-free operation (PFFO), putting in Table 2.2 the cost of different variants and in Table 2.3 the values of system PFFO.

On the basis of these two tables, one can easily compile a new table (Table 2.4) with triplets ordered by cost. One can see that there are such triplets $X^{(k)}$ and $X^{(k+1)}$ that $C(X^{(k+1)}) > C(X^{(k)})$ but

TABLE 2.3   Values of System PFFO for Various X

			x_2		
			1	2	3
			0.600	0.840	0.936
x_1	1	0.700	0.420	0.588	0.655
	2	0.910	0.546	0.764	0.852
	3	0.973	0.584	0.817	0.911
	4	0.992	0.595	0.833	0.928
	5	0.998	0.599	0.838	0.934

TABLE 2.4   List of Triplets Ordered by the System Cost

$C(X)$	$R(X)$	$X^{(j)}$	x_1	x_2	
2.7	0.420	(1)	1	1	
3.9	0.546	(2)	2	1	
5.1	0.584	(3)	3	1	
6.3	0.595	(4)	4	1	
6.6	0.588	(5)	1	2	*
7.5	0.599	(6)	5	1	
7.8	0.764	(7)	2	2	
9.0	0.817	(8)	3	2	
9.3	0.655	(9)	1	3	*
10.2	0.833	(10)	4	2	
10.5	0.852	(11)	2	3	
11.4	0.838	(12)	5	2	*
11.7	0.911	(13)	3	3	
12.9	0.928	(14)	4	3	
14.1	0.934	(15)	5	3	

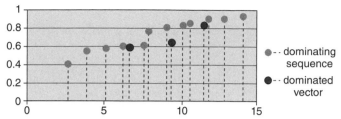

FIGURE 2.2 Dominating sequence and dominated vectors for the numerical example.

$R(X^{(k+1)}) < R(X^{(k)})$. In this case, it is said that triplet $\Delta^{(k)}$ ($X^{(k)}$) = {$X^{(k)}$, $R(X^{(k)})$, $C(X^{(k)})$} dominates over triplet $\Delta^{(k+1)}$ ($X^{(k+1)}$) = {$X^{(k+1)}$, $R(X^{(k+1)})$, $C(X^{(k+1)})$}. Such triplets are excluded in further analysis. (In Table 2.4 these vectors are $X^{(5)}$, $X^{(9)}$ $X^{(12)}$.) All remaining vectors $X^{(k)}$ are called *dominating* (see Fig. 2.2).

Based on Table 2.4, one can easily find desired solutions:

(a) For the direct optimal redundancy problem, one finds the largest value of cost that is still admissible.

(b) For the inverse optimal redundancy problem, one finds the smallest value of reliability index that exceeds the required value.

For the considered numerical example, the direct problem solution is vector $X^{(8)}$ (reliability index = 0.817 and system cost = 9) and for the inverse problem solution is vector $X^{(4)}$ (system cost = 6.3 and reliability index = 0.595).

2.3 FORMULATION OF OPTIMAL REDUNDANCY PROBLEMS WITH MULTIPLE CONSTRAINTS

2.3.1 Direct Optimal Redundancy Problem

Sometimes the optimal redundancy problem is formulated for multiple constraints, for instance, maximization of a reliability index under conditions in which other factors (cost, volume, weight, etc.) are limited by some fixed conditions. This problem can be written as:

$$\max_X \left\{ R(X) | C_1(X) \geq C_1^0, C_2(X) \geq C_2^0, \dots, C_M(X) \geq C_m^0 \right\}, \quad (2.9)$$

where $C_j^0, j = 1, 2, \dots, M$ are given constraints on the corresponding type of expenditures for the system as a whole.

In this case, further detailed considerations, such as those in the section above, are possible, though we omit them for the sake of brevity.

2.3.2 Inverse Optimal Redundancy Problem

Very rarely can one find the following problem: a system is designated for multiple tasks and performing each task requires different parts of the system. Sets of such system parts may be called subsystems. Some parts of the system are used for all tasks and some only for performance-specific tasks. Tasks of these subsystems may have different reliability requirements (for instance, some subsystems may perform extraordinary important tasks).

For such systems, one can formulate the following problem:

$$\min_X \left\{ C(X) | R_1(X) \geq R_1^0, R_2(X) \geq R_2^0, \dots, R_M(X) \geq R_m^0 \right\}. \quad (2.10)$$

To make the problem clearer, consider a simple illustrative example where a system of four units is conditionally depicted as three interdependent subsystems (Fig. 2.3).

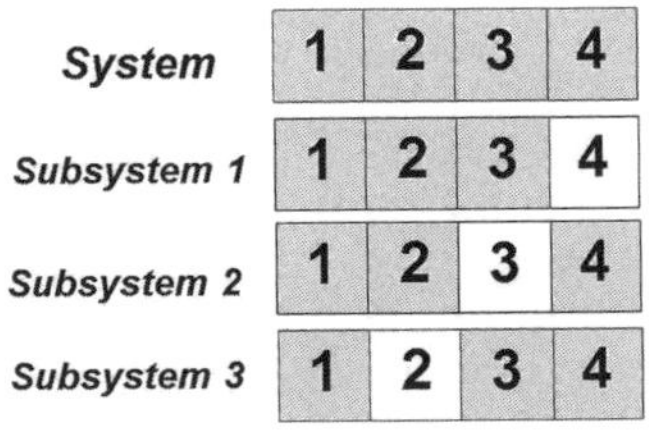

FIGURE 2.3 Conditional dividing of a system by subsystems.

For this system as a whole the inverse problem of optimal redundancy can be written as:

$$\min_{X}\{C(x)|r_1(x_1)r_2(x_2)r_3(x_3) \geq R_1^0, r_1(x_1)r_2(x_2)r_4(x_4) \geq R_2^0,$$
$$r_1(x_1)r_3(x_3)r_4(x_4) \geq R_3^0\}. \tag{2.11}$$

Example 2.2

Consider the same system as in Example 2.1. Introduce one more unit parameter, say, weight, W. Unit parameters are given in Table 2.5. We need to find a number of units of both types, $X^{opt} = (x_1^{opt}, x_2^{opt})$, that maximize system reliability index under constraints on both limiting factors: $C(X) \leq 7$ units of cost and $W(X) \leq 10$ units of weight, that is, we consider the inverse optimization problem.

For the solution, use Table 2.6 and Table 2.7 and add to them a new one for total system weight.

TABLE 2.5 Unit Parameters for Example 2.2

	PFFO	Cost	Weight
Unit 1	0.7	1.2	2.3
Unit 2	0.6	2.7	1.5

TABLE 2.6 Values of System Weight for Various X

		x_2		
		1	2	3
		1.5	3	4.5
x_1	1	2.3	5	7.7	10.4
	2	4.6	7.3	10	12.7
	3	6.9	9.6	12.3	15
	4	9.2	11.9	14.6	17.3
	5	11.5	14.2	16.9	19.6

TABLE 2.7 List of Triplets Ordered by the System Cost

$C(X)$	$W(X)$	$R(X)$	$X^{(i)}$	x_1	x_2	
2.7	5.0	0.420	(1)	1	1	
3.9	7.3	0.546	(2)	2	1	
5.1	9.6	0.584	(3)	3	1	
6.3	11.9	0.595	(4)	4	1	
6.6	7.7	0.588	(5)	1	2	optimum
7.5	14.2	0.599	(6)	5	1	
7.8	10.0	0.764	(7)	2	2	
9.0	12.3	0.817	(8)	3	2	
9.3	10.4	0.655	(9)	1	3	*
10.2	14.6	0.833	(10)	4	2	
10.5	12.7	0.852	(11)	2	3	
11.4	16.9	0.838	(12)	5	2	*
11.7	12.3	0.911	(13)	3	3	
12.9	17.3	0.928	(14)	4	3	
14.1	16.9	0.934	(15)	5	3	

On the basis of Table 2.5 and Table 2.6, let us compile a new table (now with quadruples, since we have four parameters) with triplets ordered by increase of cost.

In Table 2.7, all cells with inadmissible cost or weight are shadowed. Thus, the maximum reachable level of reliability index under the given constraints is 0.588, and it is reached by vector $X^{(5)}$. It is interesting to notice that in the previous example this vector was dominated and could not be a solution.

In case of multi-constraint situations, a dominating sequence also exists. For instance, vector $X^{(12)}$ is dominated by vector $X^{(11)}$: for larger values $C(X) = 11.4$ and $W(X) = 16.9$ the reliability index, $R(X) = 0.838$, is smaller than 0.852. Another such pair of vectors is $X^{(9)}$ and $X^{(7)}$: both parameters "cost-weight" for $X^{(9)}$ (9.3; 10.4) are correspondingly larger than analogous parameters for $X^{(7)}$, though the latter vector is characterized by a larger reliability index (all dominated vectors are marked with "*").

We don't supply a numerical example for direct problem solution due to its clumsiness.

2.4 FORMULATION OF MULTI-CRITERIA OPTIMAL REDUNDANCY PROBLEMS

2.4.1 Direct Multi-Criteria Optimal Redundancy Problem

Assume that a designer has to reach the required level of reliability having several limiting factors, such as cost, weight, and volume. Usually all these factors are somehow dependent: a miniature unit can be more expensive; weight and volume of a unit are naturally dependent; and so on. What does it mean to say "the best solution" in this case? Solutions satisfying the same reliability requirements can be incomparable: one variant has a smaller total weight, for example.

Actually, the problem of choosing a preferable solution lies outside the scope of mathematics: it is up to a decision maker. However, there are some useful procedures for finding the so-called *non-improvable solutions*. This means that none of these selected variants (solutions) is strictly better than another, but one is chosen in accordance to some subjective measures of preference.

A set of the multi-criteria problem solutions is called the *Pareto set* (see Fig. 2.4). In mathematical terms one can write the problem in the form:

$$\underset{X}{\text{MIN}}\left\{C_1(X), C_2(X), \ldots, C_M(X) | R(X) \geq R^0\right\}, \qquad (2.12)$$

where the symbol MIN in capital letters denotes Pareto "minimization" of vector $\{C_1(X), C_2(X), \ldots, C_M(X)\}$.

All Pareto solutions for condition $R(X) \geq R^0$ are dominating in a vector sense: for each Pareto-optimal vector $X^{(k)}$, there is no vector $X' < X^{(k)}$ that $R(X') \geq R(X^{(k)})$.

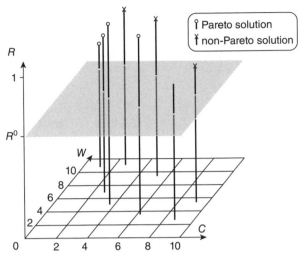

FIGURE 2.4 Explanation of the Pareto solutions for direct multi-criteria problem.

2.4.2 Inverse Multi-Criteria Optimal Redundancy Problem

The inverse problem for a multi-criteria case can be written as:

$$\text{MAX}_{X}\{R_1(X), R_2(X), \ldots, R_M(X)|C(X) \le C^0\}, \qquad (2.13)$$

where the symbol MAX in capital letters denotes Pareto "maximization" of vector $\{R_1(X), R_2(X), \ldots, R_M(X)\}$.

For instance, for the system depicted in Figure 2.3, one can write:

$$\text{MAX}_{X}\{R_1(X), R_2(X), R_3(X)|C(X) \le C^0\}. \qquad (2.14)$$

Decisions about what variant of the system configuration should be chosen have to be made by a decision maker. Expert opinion cannot be formalized; this is why experts still survive in the modern computer age!

CHRONOLOGICAL BIBLIOGRAPHY

Moskowitz, F., and Mclean, J. B. 1956. "Some reliability aspects of system design." *IRE Transactions on Reliability and Quality Control*, no. 3.

Black, G., and Proschan, F. 1959. "On optimal redundancy." *Operations Research*, no. 5.

Ghare, P. M., and Taylor, R. E. 1959. "Optimal redundancy for reliability in series systems." *Operations Research*, no. 5.

Proschan, F. 1960. "On optimal supply." *Naval Research Logistics Quarterly*, no. 4.

Morrison, D. F. 1961. "The optimum allocation of spare components in systems." *Technometrics*, vol. 5, no. 5.

Blitz, M. 1963. "Optimum allocation of spare budget." *Naval Research Logistics Quarterly*, no. 2.

Gertsbakh, I. B. 1966. "Optimum use of reserve elements." *Soviet Journal of Computer and System Sciences*, no. 5.

Lawler, E. L., and Bell, M. D. 1966. "A method for solving discrete optimization problems." *Operations Research*, no. 14.

Alekseev, O. G. 1967. "On one problem of optimal redundancy." *Engineering Cybernetics*, no. 1.

Alekseev, O.G. 1967. "Optimization of complex systems reliability under several constraints." *Automatics and Telemechanics*, no. 12.

Raikin, A.L., and Mandel, A.S. 1967. "Construction of the optimal schedule of redundant units switching." *Automation and Remote Control*, no. 5.

El-Neweihi, E., Proschan, F., and Sethuraman, J. 1968. "Optimal allocation of components in parallel-series and series-parallel systems." *Journal on Applied Probability*, no. 3.

Mizukami, K. 1968. "Optimum redundancy for maximum system reliability by the method of convex and integer programming." *Operations Research*, no. 2.

Ushakov, I.A. 1969. *Method of Solving Optimal Redundancy Problems under Constraints* (in Russian). Sovetskoe Radio.

Jensen, P. A. 1970. "Optimization of series-parallel-series networks." *Operations Research*, no. 3.

Messinger, M., and Shooman, M. L. 1970. "Technique for optimal spare allocation." *IEEE Transactions on Reliability*, no. 11.

Shaw, L., and Sinkar, S. G. 1974. "Redundant spares allocation to reduce reliability costs." *Naval Research Logistics Quarterly*, no. 2.

Agarwal, K. K., Gupta, J. S. and Misra, K. B. 1975. "A new heuristic criterion for solving a redundancy optimization problem." *IEEE Transactions on Reliability*, no. 24.

Aggarval, K. V. 1976. "Redundancy optimization in general systems." *IEEE Transactions on Reliability*, no. 5.

Nakagawa, Y., and Nakashima, K. 1977. "A heuristic method for determining reliability allocation." *IEEE Transactions on Reliability*, no. 26.

Tillman, F. A., Hwang, C. L., and Kuo, W. 1977. "Determining component reliability and redundancy for optimum system reliability." *IEEE Transactions on Reliability*, no. 26.

Kuo, W., Hwang, C. L., and Tillman, F. A. 1978. "A note on heuristic methods in optimal system reliability." *IEEE Transactions on Reliability*, no. 5.

Nakagawa, Y., and Miyazaki, S. 1981. "An experimental comparison of the heuristic methods for solving reliability optimization problems." *IEEE Transactions on Reliability*, no. 30.

Bulfin, R. L., and Liu, C. Y. 1985. "Optimal allocation of redundant components for large systems." *IEEE Transactions on Reliability*, no. 3.

Malashenko, Y. E., Shura-Bura, A. E., and Ushakov, I. A. 1985. "Optimal redundancy." In *Handbook: Reliability of Technical Systems*, I. Ushakov, ed. Sovetskoe Radio.

Dinghua, S. 1987. "A new heuristic algorithm for constrained redundancy-optimization in complex systems." *IEEE Transaction on Reliability*, no. 36.

Boland, P. J., El-Neweihi, E., and Proschan, F. 1991. "Redundancy importance and allocation of spares in coherent systems." *Journal of Statistical Planning and Inference*, no. 1–2.

Yanagi, S., Aso, K., and Sasaki, M. 1992. "Optimal spare allocation problem based on the interval estimate of availability." *International Journal of Production Economics*, no. 3.

El-Neweihi, E., and Sethuraman, J. 1993. "Optimal allocation under partial ordering of lifetimes of components." *Advances in Applied Probability*, no. 4.

Kim, J. H., and Yum, B. J. 1993. "A heuristic method for solving redundancy optimization problems in complex systems." *IEEE Transaction on Reliability*, no. 42.

Jianping, L. 1996. "A bound heuristic algorithm for solving reliability redundancy optimization." *Microelectronics and Reliability*, no. 3.

Levitin, G., Lisnianski, A., and Elmakis, D. 1997. "Structure Optimization of Power System with Different Redundant Elements." *Electric Power Systems Research*, no. 43.

Rubinstein, R. Y., Levitin, G., Lisnianski, A., and Ben Haim, H. 1997. "Redundancy optimization of static series-parallel reliability models under uncertainty." *IEEE Transactions on Reliability*, no. 4.

Prasad, V. R., and Raghavachari, M. 1998. "Optimal allocation of interchangeable components in a series-parallel system." *IEEE Transactions on Reliability*, no. 1.

Lisnianski, A., Levitin, G., Ben Haim, H., and Elmakis, D. 1999. "Power system optimization subject to reliability constraints." *Electric Power Systems Research*, no. 39.

Mi, J. 1999. "Optimal active redundancy allocation in k-out-of-n system." *Journal on Applied Probability*, no. 3.

Kuo, W., and Zuo, M. J. 2003. *Optimal Reliability Modeling: Principles and Applications*. John Wiley & Sons.

Ha, C., and Kuo, W. 2006. "Reliability redundancy allocation: An improved realization for nonconvex nonlinear programming problems." *European Journal of Operational Research*, no. 1.

Kim, H-G., Bae, C.-O., and Park, D.-J. 2006. "Reliability-redundancy optimization using simulated annealing algorithms." *Journal of Quality in Maintenance Engineering*, no. 4.

Kuo, W., Prasad, V. R., Tillman, F. A., and Hwang, C.-L. 2006. *Optimal Reliability Design: Fundamentals and Applications*. Cambridge University Press.

Dai Y., and Levitin, G. 2007. "Optimal resource allocation for maximizing performance and reliability in tree-structured grid services." *IEEE Transactions on Reliability*, no. 3.

Kuo, W., and Wan, R. 2007. "Recent advances in optimal reliability allocation." *Computational Intelligence in Reliability Engineering*, no. 1.

Liang, Y.-C., Lo, M.-H., and Chen, Y.-C. 2007. "Variable neighborhood search for redundancy allocation problems." *IMA Journal of Management Mathematics*, no. 2.

Liang, Y.-C., and Chen, Y.-C. 2007. "Redundancy allocation of series-parallel systems using a variable neighborhood search algorithm." *Reliability Engineering and System Safety*, no. 3.

Tavakkoli-Moghaddam, R. 2007. "A new mathematical model for a redundancy allocation problem with mixing components redundant and choice of redundancy strategies." *Applied Mathematical Sciences*, no. 45.

CHAPTER 3

METHOD OF LAGRANGE MULTIPLIERS

One of the first attempts to solve the optimal redundancy problem was based on the classical *Lagrange Multiplier Method*. This method was invented and developed by great French mathematician Lagrange (see Box). This method allows one to get the extreme value of the function under some specified constraint on another involved function in the form of equality. The Lagrange Multiplier Method is applicable if both functions (optimizing and constraining) are monotone and differentiable.

Optimal Resource Allocation: With Practical Statistical Applications and Theory,
First Edition. Igor A. Ushakov.
© 2013 John Wiley & Sons, Inc. Published 2013 by John Wiley & Sons, Inc.

Joseph-Louis Lagrange (1736–1813)

Lagrange made outstanding contributions to all fields of analysis, to number theory, and to classical and celestial mechanics. He was one of the creators of the calculus of variations, and also introduced the method of Lagrange multipliers where possible constraints were taken into account. He invented the method of solving differential equations known as variation of parameters and applied differential calculus to the theory of probabilities. Lagrange studied the three-body problem for the earth, sun, and moon and the movement of Jupiter's satellites. Above all, he contributed to mechanics, having transformed Newtonian mechanics into a new branch of analysis, Lagrangian mechanics.

Strictly speaking, this method is not appropriate for optimal redundancy problem solving because the system reliability and cost are described by functions of discrete arguments x_i (numbers of redundant units) and the restrictions on accessible resources (or on required values of reliability) are fixed in the form of inequalities.

Nevertheless, this method is interesting in general and also gives us some useful hints for the appropriate solution of some practical problems of a discrete nature.

Let us begin with the direct optimal redundancy problem (Fig. 3.1 below). To solve this problem, we construct the Lagrange function, $\mathcal{L}(X)$:

$$\mathcal{L}(X) = C(X) + \Lambda R(X) \tag{3.1}$$

where $C(X)$ and $R(X)$ are the cost of the system redundant units and the system reliability, respectively, if there are X redundant units of all types, $X = (x_1, x_2, \ldots, x_n)$.

The goal is to minimize $C(X)$ taking into account constraints in the form of $R(X^{opt}) = R^0$. Thus, the system of equations to be solved is

$$\begin{cases} \dfrac{\partial \mathcal{L}(X)}{\partial x_i} = \dfrac{\partial C(X)}{\partial x_i} + \Lambda \dfrac{\partial R(X)}{\partial x_i} = 0 \\ \text{for all } i = 1, 2, \ldots, n, \text{ and} \\ R(X^{opt}) = R^0. \end{cases} \tag{3.2}$$

The values to be found are: x_i^{opt}, $i = 1, 2, \ldots, n$, and Λ.

If both function $C(X)$ and $R(X)$ are separable[1] and differentiable, the first n equations of (3.2) can be rewritten as

---

[1] Taking the logarithm of multiplicative function $R(X)$, one gets an additive function of logarithms of multipliers.

$$\begin{cases} \dfrac{dL_1(x_1)}{dx_1} = \dfrac{dC_1(x_1)}{dx_1} + \Lambda \dfrac{dR_1(x_1)}{dx_1} = 0 \\[2mm] \dfrac{dL_n(x_n)}{dx_n} = \dfrac{dC_n(x_n)}{dx_n} + \Lambda \dfrac{dR_n(x_n)}{dx_n} = 0 \\[2mm] R(X^{opt}) = R^0. \end{cases} \qquad (3.3)$$

On a physical level, Equation (3.3) means that for separable functions $L(X)$ and $C(X)$, the optimal solution corresponds to equality of relative increments of reliability of each redundant group for an equal and infinitesimally small resources investment.

$$\frac{dC_1(x_1)}{dx_1} \cdot \left(\frac{dR_1(x_1)}{dx_1} \right)^{-1} = \frac{dC_2(x_2)}{dx_2} \cdot \left(\frac{dR_2(x_2)}{dx_2} \right)^{-1}$$

$$= \cdots = \frac{dC_n(x_n)}{dx_n} \cdot \left(\frac{dR_n(x_n)}{dx_n} \right)^{-1} = -\Lambda. \qquad (3.4)$$

In the general case, Equation (3.3) yields no closed form solution but it is possible to suggest an algorithm for numerical calculation:

1. At some arbitrary point $x_1^{(1)}$ calculate the derivative for some fixed redundant group, say, the first one:

$$\left. \frac{dR_1(x_1)}{c_1 dx_1} \right|_{x_1^{(1)}} = -\Lambda^{(1)}. \qquad (3.5)$$

Of course, one would like to choose a value of $x_1^{(1)}$ close to an expected optimal solution x_i^{opt}. For example, if you consider spare parts for equipment to operate failure-free during time t and know that the unit mean time to failure (MTTF) is T, this value should be a little larger than t/T. In other words, this choice should be done based on engineering experience and intuition.

2. For the remaining redundant groups, calculate derivatives until the following condition is satisfied:

$$\left.\frac{dR_i(x_i)}{c_i dx_i}\right|_{x_i^{(1)}} = -\Lambda^{(1)}. \tag{3.6}$$

3. Calculate value:

$$R(X^{(1)}) = \sum_{1 \le i \le n} R_i(x_i^{(1)}). \tag{3.7}$$

4. Compare $R^{(1)}$ with R^0. If $R^{(1)} > R^0$, choose $x_i^{(2)} < x_i^{(1)}$; if $R^{(1)} < R^0$, choose $x_i^{(2)} > x_i^{(1)}$. After choosing a new value of $x_i^{(2)}$, return to step 1 of the algorithm.

The stopping rule: $X^{(N)}$ is accepted as the solution if the following condition holds:

$$\left|R(X^{(N)}) - R^0\right| \le \varepsilon, \tag{3.8}$$

where ε is some specified admissible discrepancy in the final value of the objective function $R(X)$.

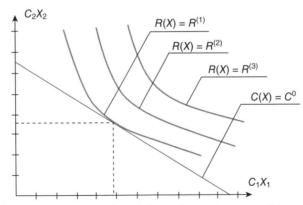

FIGURE 3.1 Procedure of solving the direct problem of optimal redundancy.

Solution of the inverse optimal redundancy problem (Fig. 3.2) is analogous. The Lagrange function is:

$$\mathcal{L}^*(X) = R(X) + \Lambda^* C(X) \tag{3.9}$$

and the equations are:

$$\begin{cases} \dfrac{\partial \mathcal{L}(X)}{\partial x_i} = \dfrac{\partial R(X)}{\partial x_i} + \Lambda \dfrac{\partial C(X)}{\partial x_i} = 0 \\ \text{for all } i = 1, 2, \ldots, n, \text{ and} \\ C(X^{opt}) = C^0. \end{cases} \tag{3.10}$$

The optimal solution has to be found with respect to goal function $C(X)$. The stopping rule in this case is

$$\left| C(X^{(N)}) - C^0 \right| \leq \varepsilon^*, \tag{3.11}$$

where ε^* is some specified admissible discrepancy in the final value of the goal function $C(X)$.

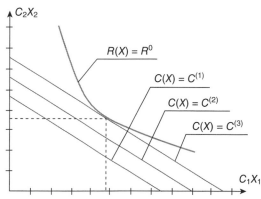

FIGURE 3.2 Procedure of solving the inverse problem of optimal redundancy.

Unfortunately, there is only one case where the direct optimal redundancy problem can be solved in a closed form. This is the case of a highly reliable system with active redundancy where:

$$R(X) = \prod_{1 \le i \le n}(1 - q_i^{x_i}) \approx 1 - \sum_{1 \le i \le n} q_i^{x_i} \qquad (3.12)$$

or

$$Q(X) \approx \sum_{1 \le i \le n} q_i^{x_i}, \qquad (3.13)$$

that is, instead of maximizing $R(X)$, one can minimize $Q(X)$ and find the needed optimal solution.

Taking into account that both objective functions are separable, Equation (3.10) can be written as:

$$\begin{cases} \dfrac{dq_i^{x_i}}{dx_i} + \Lambda c_i = 0 \\ \text{for all } i = 1, 2, \ldots, n, \text{ and} \\ Q(X^{opt}) = Q^0. \end{cases} \qquad (3.14)$$

From equation

$$\frac{dq_i^{x_i}}{dx_i} = \ln q_i \cdot q_i^{x_i} \qquad (3.15)$$

we can finally write

$$q_j^{x_j} = \frac{c_j}{\Lambda(-\ln q_j)}. \qquad (3.16)$$

Now returning to the last equation in Equation (3.10), we have

$$Q^0 = \frac{1}{\Lambda} \sum \frac{c_j}{(-\ln q_j)}, \qquad (3.17)$$

and, consequently, the Lagrange multiplier has the form

$$\Lambda = \frac{1}{Q^0} \sum \frac{c_j}{(-\ln q_j)}. \qquad (3.18)$$

From Equation (3.16), one has

$$x_j = \frac{1}{\ln q_i} \cdot \ln\left[\frac{c_j}{\Lambda(-\ln q_j)}\right]. \tag{3.19}$$

Finally, after substitution of Equation (3.17) into Equation (3.19), one gets

$$x_j = \frac{1}{\ln q_i} \cdot \ln\left[\frac{c_j}{\Lambda(-\ln q_j)}\right] \cdot \left(\frac{1}{Q^0}\sum \frac{c_j}{(-\ln q_j)}\right)^{-1}. \tag{3.20}$$

Hardly anybody would be happy to deal with such a clumsy formula!

The solutions obtained by the Lagrange Multiplier Method are usually continued due to requirements to the objective functions. Immediate questions arise: Is it possible to use an integer extrapolation for each non-integer x_i? If this extrapolation is possible, is the obtained solution optimal?

Unfortunately, even if one tries to "correct" non-integer solutions by substituting lower and upper integer limits $\underline{x_j} < x_j < \overline{x_j}$, this very rarely leads to an optimal solution! Moreover, enumerating all 2^n possible "corrections" can itself be a problem. We demonstrate this statement on a simple example.

CHRONOLOGICAL BIBLIOGRAPHY

Everett, H. 1963. "Generalized Lagrange multiplier method for solving problems of optimum allocation of resources." *Operations Research*, no. 3.

Greenberg, H. J. 1970. "An application of a Lagrangian penalty function to obtain optimal redundancy." *Technometrics*, no. 3.

Hwang, C. L., Tillman, F. A., and Kuo, W. 1979. "Reliability optimization by generalized Lagrangian-function and reduced-gradient methods." *IEEE Transactions on Reliability*, no. 28.

Kuo, W., Lin, H. H., Xu, Z., and Zhang, W. 1987. "Reliability optimization with the Lagrange-multiplier and branch-and-bound technique." *IEEE Transactions on Reliability*, no. 5.

CHAPTER 4

STEEPEST DESCENT METHOD

4.1 THE MAIN IDEA OF SDM

The *Steepest Descent Method* (*SDM*) is based on a very natural idea: moving from an arbitrary point in the direction of the maximal gradient of the goal function, it is possible to reach the maximum of a multi-dimensional unimodal function. The origin of the method's name lies in the fact that a water drop runs down a non-flat surface choosing the direction of instantaneous maximum descent.

The next simple example explains the algorithm more graphically. Suppose that a traveler comes to a hill that is hidden in a thick mist. His goal is to reach the hill top with no knowledge about the mountain shape except the fact that the hill is smooth enough (has no ravines or local hills). The traveler sees only a very restricted area around the starting point. The question is: What is the shortest

Optimal Resource Allocation: With Practical Statistical Applications and Theory, First Edition. Igor A. Ushakov.
© 2013 John Wiley & Sons, Inc. Published 2013 by John Wiley & Sons, Inc.

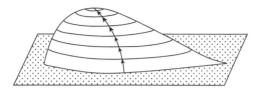

FIGURE 4.1   A path of a traveler up to the hill top.

path from the initial point to the mountain's top? Intuition hints that the traveler has to move in the direction of the maximal possible ascent at each point on his path to the mountain's top (Fig. 4.1). This direction coincides with the gradient of the function at each point.

However, optimal redundancy problems have an integer nature: redundant units can be added to the system one by one. The previous analogy is useful in case of continuous functions of continuous arguments. But in the case of optimal redundancy, all arguments are discrete. If we continue the analogy with the traveler, one sees that there are restrictions on the traveler's movement: he can move only in the north–south or east–west directions and can change the direction only at the vertices of a discrete grid with specified steps (Fig. 4.2). This means that at each vertex one should use the direction of the largest partial derivative. Because of this, one sometimes speaks of the *Method of Coordinate Steepest Descent.*

This idea of finding the maximum of a unimodal function may be applied to the optimal redundancy problem.

4.2   DESCRIPTION OF THE ALGORITHM

Consider a system that is a series connection of independent redundant groups. Let us use the both goal functions in the additive form

$$L(X) = \ln R(X) = \sum_{1 \le i \le n} \ln R_i(x_i) \tag{4.1}$$

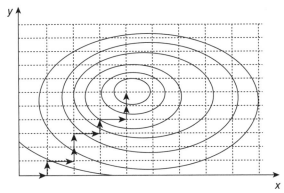

FIGURE 4.2 A traveler's path when only north–south and east–west directions are allowed.

and

$$C(X) = \sum_{1 \leq i \leq n} c_i x_i. \tag{4.2}$$

It is clear that maximization of function $R(X)$ corresponds to minimization of function $L(X)$, that is, the optimum solution X^{opt} for goal function $L(X)$ delivers as well the optimum for function $R(X)$.

    Introduce vector $X^{(N)} = (x_1^{(N)}, \ldots, x_n^{(N)})$, where $x_i^{(N)}$ is the number of redundant units of the ith redundant group at the nth step of the SDM process. Denote by $R_i(x_i^{(N)})$ the reliability index and by $C_i(x_i^{(N)})$ the cost of the ith redundant group after the nth step of the SDM process. For the convenience of further exposition, let us also introduce the following additional notation: $X_i^{(N)} = (x_1^{(N)}, \ldots, x_{i-1}^{(N)}, 0, x_{i+1}^{(N)}, \ldots, x_n^{(N)})$, that is, vector $X_i^{(N)}$ is vector X without component x_i. Obviously, $X^{(N)} = X_i^{(N)} + x_i^{(N)}$.

    At the nth step of the process, one adds a redundant unit of type k, for which the relative increase of reliability index is at the maximum, that is,

$$\gamma_k^{(N)}(x_i^{(N)}) = \max_{1 \le i \le n} \frac{\ln R(x_i^{(N)}) - \ln R(x_i^{(N)} + 1)}{c_i}. \qquad (4.3)$$

A unit of this type is added to the set of the system's redundant units. The process continues in the same manner until the optimal solution is obtained.

Now let us describe the optimization algorithm step by step from the very beginning.

1. Before the beginning of the process, there are no redundant units at all, that is, $x_1^{(0)} = \cdots = x_n^{(0)}$ or, in other words, $X^{(0)} = \vec{0}$.
2. For all i, $i=1, 2, \ldots, n$, one calculates values $\gamma_i^{(0)}(x_i^{(0)})$.
3. One finds such index k that delivers maximum

$$\gamma_k^{(0)}(x_k^{(0)}) = \max_{1 \le i \le n} \gamma_i^{(0)}(x_i^{(0)}).$$

4. One calculates a new value

$$x_k^{(1)} = x_k^{(0)} + 1.$$

5. All other $x_i^{(0)}$, $i \ne k$, change their superscripts: $x_i^{(0)} \Rightarrow x_i^{(1)}$.
6. One gets a new vector of the system's redundant units:

$$X^{(1)} = (X_k^{(0)} + x_k^{(1)}).$$

7. One calculates the value of $L(X^{(1)})$ and determines the corresponding $R(X^{(1)})$.
8. One calculates the value of $L(X^{(1)})$.
9. One calculates by the same rule a new value $\gamma_k^{(1)}(x_k^{(1)})$.
10. All other values $\gamma_i^{(0)}(x_i^{(0)})$, $i \ne k$, are conserved but one changes their superscripts:

$$\gamma_i^{(0)}(x_i^{(0)}) \Rightarrow \gamma_i^{(1)}(x_i^{(1)}).$$

11. GOTO (3).

4.3 THE STOPPING RULE

The solution of the direct problem of optimal redundancy is reached at such step N for which the following condition is valid:

$$C(X^{(N)}) \leq C^0 < C(X^{(N+1)}). \tag{4.4}$$

The value of $R(X^{(N)})$ is the maximum possible for the given constrain on the system cost.

Sometimes the SDM procedure requires one to add at the last step a very expensive unit but adding this unit exceeds the given constraint on the total cost of the system's redundant units. At the same time, if one does not add this unit, there are some extra financial resources to add other, less expensive units. In this case one may bypass the expensive unit and continue the procedure.

The inverse optimal redundancy problem reaches its optimal solution at step N where the following condition is valid:

$$R(X^{(N-1)}) < R^0 \leq R(X^{(N)}). \tag{4.5}$$

The value of $C(X^{(N)})$ is the minimum possible for the given constraint on the required system reliability index.

American mathematician Frank Proschan (Barlow and Proschan, 1965, 1981; see Box) has proven that the SDM procedure delivers members of dominating sequences if each function $R_i(x_i)$ is concave.

Frank Proschan (1921–1993)

Frank Proschan was an American mathematician who earned his Ph.D. in statistics from Stanford University in 1959. He had held positions with the Federal Government at the National Bureau of Standards (1941–1952), with Sylvania Electric Products (1952–1960), and with Boeing Scientific Labs (1960–1970). Beginning in 1970, he was Professor of Statistics at Florida State University. He won many honors including the Von Neumann Prize award presented by TIMS-ORSA. He was a Fellow of the Institute of the American Statistical Association and a member of the International Statistical Institute.

TABLE 4.1   Units' Parameters

Unit type	Number of units, n_k	Unit failure rate, λ_k $(10^{-5} \text{ hr}^{-1})$	Unit cost, c_k
1	5	1	1
2	10	1	1
3	5	1	8
4	10	1	8
5	5	8	1
6	10	8	1

Example 4.1

A series system consists of six different units whose parameters are given in Table 4.1. (Distributions of time to failure are assumed exponential.)

One needs to find the optimum number of standby units for successful system operation during $t_0 = 1000$ hours for two cases:

(1) Required probability of failure-free operation (PFFO) is 0.9995.

(2) Admissible expenses on all spare units are 40 cost units.

First, using Table 4.1, let us find parameters of Poisson distributions for each group by the formula $a_i = \lambda_i n_i t_0$. The probability of appearance of exactly k failures during given time t_0 is calculated by the formula:

$$q_i(k) = \frac{(\lambda_i n_i t_0)^k}{k!} \exp(-\lambda_i n_i t_0). \tag{4.6}$$

By the way, if condition $Q_i(x_i) \gg Q_i(x_i + 1)$ holds, then calculation of values γ can be simplified up to

TABLE 4.2 Values of Parameters of Poisson Distribution

Parameter	Value
a_1	$1 \cdot 10^{-5} \cdot 5 \cdot 1000 = 0.05$
a_2	$1 \cdot 10^{-5} \cdot 10 \cdot 1000 = 0.1$
a_3	$1 \cdot 10^{-5} \cdot 5 \cdot 1000 = 0.05$
a_4	$1 \cdot 10^{-5} \cdot 10 \cdot 1000 = 0.1$
a_5	$8 \cdot 10^{-5} \cdot 5 \cdot 1000 = 0.4$
a_6	$1 \cdot 10^{-5} \cdot 10 \cdot 1000 = 0.8$

TABLE 4.3 Values of Unreliability Indices for Various x_i

x_i	$q_1(x_1)$	$q_2(x_2)$	$q_3(x_3)$	$q_4(x_4)$	$q_5(x_5)$	$q_6(x_6)$
0	0.0476	0.0905	0.0476	0.0905	0.268	0.359
1	0.00119	0.00452	0.00119	0.00452	0.0536	0.144
2	1.98E–05	0.000151	1.98E–05	0.000151	0.00715	0.0383
3	2.48E–07	3.77E–06	2.48E–07	3.77E–06	0.000715	0.00767
4	2.48E–09	7.54E–08	2.48E–09	7.54E–08	5.72E–05	0.00123
5	2.06E–11	1.26E–09	2.06E–11	1.26E–09	3.81E–06	0.000164
6	1.47E–13	1.8E–11	1.47E–13	1.8E–11	2.18E–07	1.87E–05
7	9.22E–16	2.24E–13	9.22E–16	2.24E–13	1.09E–08	1.87E–06
8	5.12E–18	2.49E–15	5.12E–18	2.49E–15	4.84E–10	1.66E–07
9	2.56E–20	2.49E–17	2.56E–20	2.49E–17	1.94E–11	1.33E–08

$$\gamma_i(x_i) \approx \frac{q_i(x_i)}{c_i}. \tag{4.7}$$

Calculated values of $q_i(x_i)$ are presented in Table 4.3.

Now we can build Table 4.4, where values $\gamma_i(x_i)$ are presented. In this table, the numbers in the upper right corner of cells are numbers corresponding to steps of the SDM procedure.

TABLE 4.4   **Values $\gamma_i(x_i)$ for All Redundant Groups**

x_i	$\gamma_1(x_1)$	$\gamma_2(x_2)$	$\gamma_3(x_3)$	$\gamma_4(x_4)$	$\gamma_5(x_5)$	$\gamma_6(x_6)$
1	2 0.0464	1 0.086	8 0.0058	6 0.0107	4 0.0268	3 0.027
2	11 0.00117	9 0.00437	16 0.000146	14 0.000547	7 0.00581	5 0.0132
3	19 1.96E–05	15 0.000147	24 2.45E–06	20 1.84E–05	13 0.000804	10 0.003834
4	28 2.45E–07	23 3.69E–06	3.07E–08	26 4.62E–07	18 8.22E–05	12 0.000805
5	2.46E–09	7.41E–08	3.07E–10	9.27E–09	22 6.67E–06	17 0.000133
6	2.05E–11	1.24E–09	2.56E–12	1.55E–10	27 4.49E–07	21 1.81E–05
7	1.47E–13	1.77E–11	1.83E–14	2.22E–12	2.59E–08	25 2.10E–06
8	9.16E–16	2.22E–13	1.15E–16	2.77E–14	1.30E–09	29 2.13E–07
9	5.09E–18	2.47E–15	6.37E–19	3.09E–16	5.81E–11	1.91E–08
	. . .	. . .	. . .	. . .	. . .	. . .

On the basis of Table 4.4, let us build the final table (Table 4.5), from which one can get needed optimal solutions. Table 4.5 allows finding optimal solutions for both optimal problems: direct as well as inverse.

By conditions of the illustrative problem, $R^0 = 0.9995$. From Table 4.5, we find that the solution is reached at step 18: unreliability in this case is 4.16E–04, that is, $R(X^{opt}) = 0.999584$ that satisfies the requirements. (Corresponding cost is 46 units.)

At the same time, for the inverse problem the solution is reached at step 15 with the total cost of 36 units. (Corresponding PFFO = 0.99669.) In this case we keep extra 4 cost units. Of course, they could be spent for 4 additional inexpensive units of types 1, 2, 5, and 6, that is, instead of the obtained solution ($x_1 = 2$, $x_1 = 3$, $x_1 = 1$, $x_1 = 2$, $x_1 = 3$, $x_1 = 4$), take solution with all resources spent: ($x_1 = 3$,

TABLE 4.5   Step-by-Step Results of SDM Procedure

	$C(X)$	$Q(X)$	x_1	x_2	x_3	x_4	x_5	x_6
...	...	...	...	...	...	...	...	...
10	24	0.021871	1	2	1	1	2	3
11	25	2.07E–02	2	2	1	1	2	3
12	26	1.43E–02	2	2	1	1	2	4
13	27	7.83E–03	2	2	1	1	3	4
14	35	3.46E–03	2	2	1	2	3	4
15	36	3.31E–03	2	3	1	2	3	4
16	44	2.14E–03	2	3	2	2	3	4
17	45	1.07E–03	2	3	2	2	3	5
18	46	4.16E–04	2	3	2	2	4	5
19	47	3.96E–04	3	3	2	2	4	5
20	55	2.49E–04	3	3	2	3	4	5
21	56	1.03E–04	3	3	2	3	4	6
22	57	5.01E–05	3	3	2	3	5	6
23	58	4.64E–05	3	4	2	3	5	6
24	66	2.69E–05	3	4	3	3	5	6
25	67	1.00E–05	3	4	3	3	5	7
26	75	6.33E–06	3	4	3	4	5	7
...	...	...	...	...	...	...	...	...

$x_1 = 4$, $x_1 = 1$, $x_1 = 2$, $x_1 = 4$, $x_1 = 5$). In this case, the system PFFO is equal to 0.99844. This solution is admissible in sense of the total cost of redundant units.

By the way, to get such a solution we could slightly change the SDM algorithm: If on a current step of the SDM procedure we "jump" over the admissible cost, we can take another unit or units with admissible cost.

Analogous corrections could be performed if the obtained current solution for the direct problem of optimal redundancy overexceeds the required value R^0.

4.5 APPROXIMATE SOLUTION

For practical purposes, an engineer sometimes needs to know an approximate solution that would be close to an optimal one a priori. Such a solution can be used as the starting point for the SDM calculation procedure. (Moreover, sometimes the approximate solution is a good pragmatic solution if input statistical data are too vague. Indeed, attempts to use strong methods with unreliable input data might be considered lacking in common sense! Remember: the "garbage-in-garbage-out" rule is valid for precise mathematical models as well!)

The proposed approximate solution (Ushakov, 1965) is satisfactory for highly reliable systems. This means that in the direct optimal redundancy problem, the value of Q^0 is very small. In a sense, such a condition is not a serious practical restriction. Indeed, if the investigated system is too unreliable, one should question if it is reasonable to improve its reliability at all. Maybe it is easier to find another solution using another system.

For a highly reliable system, one can write:

$$Q^0 \approx Q(X^{(N)}) \approx \sum_{1 \le i \le n} Q_i(x_i^{(N)}) \tag{4.8}$$

at the stopping moment (the nth step) of the optimization process, when the value of the reliability index should be high enough.

From Table 4.4, one can see that there is a "strip" that divides all values of $\gamma_i(x_i)$. For instance, consider the cells corresponding to steps 19–24 (shadowed on the table). The largest value laying above this strip (step 18) has value $\gamma = 8.22E{-}05$, which is larger than any value of γ belonging to the strip. At the same time, the largest value laying below the strip (step 25) has value $\gamma = 2 \cdot 10^{-6}$, which is smaller than any corresponding value on the strip. This means that there is some value Λ, $2 \cdot 10^{-6} < \Lambda < 8.22 \cdot 10^{-5}$, that divides all sets of γ into two specific subsets: this Λ, in a sense, plays the role of a Lagrange multiplier. Indeed, the approximate equality of γ for each "argument" x_i completely corresponds to the equilibrium in the Lagrange solution.

Let us make a reasonable assumption that at the stopping moment

$$\gamma_1^{(N)} \approx \gamma_2^{(N)} \approx \cdots \approx \gamma_n^{(N)} \approx \Lambda. \tag{4.9}$$

At the same time,

$$\gamma_i^{(N)} \approx \frac{Q_i(x_i^{(N)})}{c_i} \approx \Lambda. \tag{4.10}$$

Now using (4.8)

$$Q^0 \approx \Lambda \sum_{1 \le i \le n} c_i, \tag{4.11}$$

and finally

$$\Lambda \approx \frac{Q^0}{\displaystyle\sum_{1 \le i \le n} c_i}. \tag{4.12}$$

Now we can substitute Equation (4.12) into Equation (4.10) and obtain:

$$Q_i(x_i^{(N)}) \approx \frac{c_i Q^0}{\displaystyle\sum_{1 \le i \le n} c_i}. \tag{4.13}$$

For solving the inverse optimal redundancy problem, one has to use a very simple iterative procedure.

(1) Find approximate starting values of the x_is

$$x_1^{(1)} = x_2^{(1)} = \cdots = x_n^{(1)} = \frac{C^0}{\displaystyle\sum_{1 \le i \le n} c_i}. \tag{4.14}$$

(2) Use these x_is to calculate $Q^{(1)}$ as

$$Q^{(1)} = \sum_{1 \le i \le n} Q_i(x_i^{(1)}). \tag{4.15}$$

(3) Calculate $g^{(1)}$ as

$$\gamma^{(1)} = \frac{Q^{(1)}}{\displaystyle\sum_{1 \le i \le n} c_i}. \tag{4.16}$$

(4) For $\gamma^{(1)}$ determine $x_i^{(2)}$ for all i from the equations

$$Q_i(x_i^{(2)}) = c_i \gamma^{(1)}. \tag{4.17}$$

(5) For all obtained $x_i^{(2)}$, one calculates the total cost of the system's redundant units as:

$$C^{(2)} = \sum_{1 \le i \le n} c_i x_i^{(2)}. \tag{4.18}$$

If $C^{(1)} > C^0$, one sets a new $\gamma^{(2)} > \gamma^{(1)}$; if $C^{(1)} < C^0$, one sets a new $\gamma^{(2)} > \gamma^{(1)}$.

After this, the procedure continues from the third step. Such an iterative procedure continues until the appropriate value of the total cost of redundant units is achieved.

CHRONOLOGICAL BIBLIOGRAPHY

Barlow, R. E., and Proschan, F. 1965. *Mathematical Theory of Reliability*. John Wiley & Sons.

Ushakov, I. A. 1965. "Approximate solution of optimal redundancy problem." *Radiotechnika*, no. 12.

Ushakov, I. A. 1967. "On optimal redundancy problems with a non-multiplicative objective function." *Automation and Remote Control*, no. 3.

Ushakov, I. A. 1969. *Method of Solving Optimal Redundancy Problems under Constraints* (in Russian). Sovetskoe Radio.

Ushakov, I. A. 1981. "Methods of Approximate Solution of Dynamic Standby Problems." *Engineering Cybernetics*, vol. 19, no. 2.

CHAPTER 5

DYNAMIC PROGRAMMING

As mentioned earlier, the problem has an essentially discrete nature, so the SDM cannot guarantee the accuracy of the solution. Thus, if an exact solution of the optimal redundancy problem is needed, one generally needs to use the *Dynamic Programming Method (DPM)*.

5.1 BELLMAN'S ALGORITHM

The main ideas of the DPM were formulated by an American mathematician Richard Bellman (Bellman, 1957; see Box), who has formulated the so-called optimality principle.

Optimal Resource Allocation: With Practical Statistical Applications and Theory, First Edition. Igor A. Ushakov.
© 2013 John Wiley & Sons, Inc. Published 2013 by John Wiley & Sons, Inc.

Richard Ernest Bellman (1920–1984)

American applied mathematician, who is famous for his invention of dynamic programming in 1953. He also made many important contributions in other fields of mathematics. Over the course of his career Bellman published 619 papers and 39 books. During the last 11 years of his life he published over 100 papers, despite suffering from the crippling complications of brain surgery. Bellman's fundamental contributions to science and engineering won him many honors, including the First Norbert Wiener Prize in Applied Mathematics (1970).

The DPM provides an exact solution of discrete optimization problems. In fact, it is a well-organized method of direct enumeration. For the accuracy of the solutions, one has to pay with a high calculation time and a huge computer memory if the problem is highly dimensional.

To solve the direct optimal redundancy problem, let us construct a sequence of Bellman's function, $B_k(r)$. This function reflects the optimal value of the goal function for a system of k redundant groups and a specified restriction r. As usual, we start at the beginning:

$$B_1(r^{(1)}) = \min_{x_1} \left\{ c_1 x_1 \,\middle|\, R_1(x_1) \ge r^{(1)}; 0 < r^{(1)} \le R^0 \right\}. \tag{5.1}$$

It is clear that in such a way we determine the number of units that are necessary for the redundant group to have a reliability index equal to $r^{(1)}$ that is within the interval $[0, R^0]$. (See Fig. 5.1.)

Now we compose the next function (see Fig. 5.2):

$$B_2(r^{(2)}) = \min_{x_2} \left\{ c_2 x_2 + B_1(r^{(1)}) \,\middle|\, r^{(1)} \cdot R_2(x_2) \ge R^0 \right\}. \tag{5.2}$$

In a sense, we have a "convolution" of the first and second redundant groups, and for each level of current redundancy, the best variant of such convolution is kept for the next stage of the procedure. In an analogous way, the recurrent procedure continues until the last Bellman's equation is compiled:

$$B_n(r^{(n)}) = \min_{x_n} \left\{ c_n x_n + B_{n-1}(r^{(n-1)}) \,\middle|\, r^{(n-1)} \cdot R_n(x_n) \ge R^0 \right\}. \tag{5.3}$$

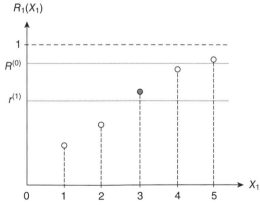

FIGURE 5.1 Illustration of the solution for Equation (5.1).

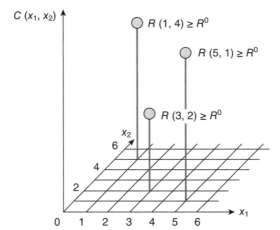

FIGURE 5.2 Illustration of the solution for Equation (5.2).

Actually, Equation (5.3) gives us only a solution for x_n: other x_is are "hidden" in previous stages of compiling Bellman's equation. Indeed, Equation (5.3) contains $B_{n-1}(r^{(n-1)})$, which allows us to determine x_{n-1}, and so on. The last found will be x_1. In a sense, the process of finding optimal x_is is going backward relating to the process of Bellman's function compiling.

Solution of the inverse problem of optimal redundancy is similar. The only difference is that the system reliability becomes an objective function and the total system cost becomes the constraint. The procedure does not need additional explanations.

At the first stage of the recurrent procedure, one compiles the Bellman's equation of the form:

$$\tilde{B}_1(c^{(1)}) = \max_{x_1}\left\{R_1(x_1)\big|c_1x_1 \le c^{(1)}; 0 < c^{(1)} \le C^0\right\}. \tag{5.4}$$

Then consequently other equations:

$$\tilde{B}_2(c^{(2)}) = \max_{x_2}\left\{R_2(x_2)\cdot\tilde{B}_1(c^{(1)})\big|c_2x_2 + c^{(1)} \le C^0\right\} \tag{5.5}$$

$$\tilde{B}_n(c^{(n)}) = \max_{x_n}\left\{R_n(x_n)\cdot\tilde{B}_{n-1}(c^{(n-1)})\big|c_nx_n + c^{(n-1)} \le C^0\right\}. \tag{5.6}$$

The optimal solution is found by the same return to the beginning of the recurrent procedure.

For illustration of the calculated procedure of dynamic programming, let us consider a simple illustrative example.

Example 5.1

Let us consider a very simple series system consisting of three units with the characteristics: $p_1 = 0.7$, $p_2 = 0.8$, $p_3 = 0.9$, and $c_1 = c_2 = c_3 = 1$. For reliability increase, a "hot" redundancy is used.

The problem is to find the optimal vector of redundant units for the system under constraint $C(X) \leq 6$ cost units.

All possible convolutions of redundant groups 1 and 2 are presented in Table 5.1. Dominating vectors within each are marked with the symbol y.

Now we compile a table (Table 5.2) with only a dominating sequence for convolution of redundant groups 1 and 2, denoting each pair as y_1, y_2, and so on.

Thus, the solution is (x_3, y_5). Now return to Table 5.1 and find there that y_5 corresponds to $x_1 = 3$ and $x_2 = 2$. This is the final step of the solving procedure.

The solution of the direct problem of optimal redundancy is more complicated for manual calculations, although it can be easily programmed for a computer.

5.2 KETTELLE'S ALGORITHM

DPM is a well-organized enumeration using convolutions of a set of possible solutions. It has some "psychological" deficiency: a researcher gets the final results without "submerging" into the solving process. If a researcher is not satisfied by a particular solution for some specified restrictions and decides to change them, it may lead to a complete resolving of the problem.

TABLE 5.1 Convolution of Redundant Groups 1 and 2

Cost	x_1	x_2	$R(x_1, x_2)$	Chosen
0	0	0	0.56	y_0
1	1	0	0.728	y_1
	0	1	0.672	
2	2	0	0.7784	
	1	1	0.8736	y_2
	0	2	0.6944	
3	3	0	0.79352	
	2	1	0.93408	y_3
	1	2	0.90272	
	0	3	0.69888	
4	4	0	0.798056	
	3	1	0.952224	
	2	2	0.965216	y_4
	1	3	0.908544	
	0	4	0.699776	
5	5	0	0.799417	
	4	1	0.957667	
	3	2	0.983965	y_5
	2	3	0.971443	
	1	4	0.909709	
	0	5	0.699955	
6	6	0	0.799825	
	5	1	0.9593	
	4	2	0.989589	
	3	3	0.990313	y_6
	2	4	0.972689	
	1	5	0.909942	
	0	6	0.699991	
. . .	. . .	. . .	. . .	. . .

TABLE 5.2

y_k	$R(y_k)$	x_3	$R_3(x_3)$	R_{syst}	Chosen
y_0	0.56	6	≈ 1	0.56	
y_1	0.728	5	0.999999	0.727999	
y_2	0.8736	4	0.99999	0.873591	
y_3	0.93408	3	0.9999	0.933987	
y_4	0.965216	2	0.999	0.964251	
y_5	0.983965	1	0.99	0.974125	X
y_6	0.990313	0	0.9	0.891282	

John D. Kettelle, Jr. (1925–2012)

John Kettelle was an American mathematician who fought for 3 years in WWII in the US Navy and then served 2 years on a submarine during the Korean War. In the next 5 years he worked in an operations research group at Arthur D. Little Co. with the founder of operations research, George Kimbell. Later, he started a series of consulting corporations. He was the author of a well-known paper on a modified dynamic programming method and edited 11 books published by ORSA. Before his death, he developed a method of negotiations using a computer as the third party.

For most practical engineering problems, using *Kettelle's Algorithm* (Kettelle, 1962; see Box) is actually a modification of the DPM. It differs from DPM by a simple and intuitively clear organization of calculating process. This algorithm is very effective for the exact solution of engineering problems due to its clarity and flexibility of calculations.

Of course, Kettelle's Algorithm, as well as DPM, requires more computer time and memory than SDM, but it gives strict solutions. At the same time, this algorithm allows one to construct an entire dominating sequence (as SDM) that allows one to switch from solving the direct optimal redundancy problem to the inverse one using the previously calculated sequence.

5.2.1 General Description of the Method

We shall describe Kettelle's Algorithm step by step.

(1) For each ith redundant group, one constructs a table of values of $R_i(x_i)$, accompanied by corresponding cost $C_i(x_i)$ (as in Table 5.3).

The sequence of these pairs for each group forms a *dominating sequence*, that is, for any j and k: $R(k) < R(k + 1)$ and $C(k) < C(k + 1)$.

TABLE 5.3 Initial Dominating Sequences for Redundant Groups

Group number	Number of redundant units in the group					
	0	1	2	. . .	n	. . .
1	$R_1(0), C_1(0),$	$R_1(1), C_1(1),$	$R_1(2), C_1(2),$	. . .	$R_1(0), C_1(0),$	. . .
2	$R_2(0), C_2(0)$	$R_2(1), C_2(1)$	$R_2(2), C_2(2)$	. . .	$R_2(0), C_2(0)$	. . .
. . .	. . .	. . .	. . .	. . .	. . .	. . .
N	$R_N(0), C_N(0)$	$R_N(1), C_N(1)$	$R_N(2), C_N(2)$	. . .	$R_N(0), C_N(0)$	. . .

(2) Take any two redundant groups from Table 5.3, say, 1 and 2, and construct compositions of pairs located in the corresponding cells by the rule: $R_{12}(x_1, x_2) = R_1(x_1) \times R_2(x_2)$ and $C_{12}(x_1, x_2) = C_1(x_1) + C_2(x_2)$ (Table 5.4).

The size of the table (i.e., values m and n) is not fixed a priori. It could be increased if a sequence of dominating pairs $\{R_{12}(x_1, x_2), C_{12}(x_1, x_2)\}$ does not include a desired solution.

As the result, now we have a system with $n - 1$ redundant groups: groups from 3 to N and one new group formed by the described composing of groups 1 and 2.

The procedure continues until one obtains a single composed group that is used in both cases: for solving direct as well as inverse problems of optimal redundancy.

5.2.2 Numerical Example

For demonstration of the Kettelle's Algorithm, let us consider a simple numerical example with a system of three redundant groups of active units (see Fig. 5.3).

Let $R_j(x_j) = 1 - q_j^{x_j}$ and $C_j(x_j) = c_j x_j$ where $1 \le j \le 3$. Assume that $q_1 = 0.3$, $q_2 = 0.5$, $q_3 = 0.5$, and $c_1 = 1$, $c_2 = 3$, $c_3 = 1$.

For the sake of calculating convenience, let us prepare in advance dominated sequences for each redundant group, presented in Table 5.5. (Notice that for a single group, sequence of a pair's "reliability-cost" is always dominating, since each added unit increases simultaneously both the cost and reliability index.)

Looking at Figure 5.4, it is easy to see that each function $R(C)$ is concave.

The next step is construction of a table (Table 5.6) with various combinations of possible configurations of groups 1 and 2 by the rules $R(x_1, x_2) = (1 - q_1^{x_1}) \times (1 - q_2^{x_2})$ and $C(x_1, x_2) = c_1 x_1 + c_2 x_2$. (Numbers in **bold italic** denote the members of dominating pairs by ascending ordering by weights.)

TABLE 5.4 Dominating Sequence for the Composition of Groups 1 and 2

Number of redundant units of the 1st redundant group		Number of redundant units of the 1st redundant group					
		0	1	2	...	n	...
	0	$R_{12}(0,0)$, $C_{12}(0,0)$	$R_{12}(1,0)$, $C_{12}(1,0)$	$R_{12}(2,0)$, $C_{12}(2,0)$	...	$R_{12}(n,0)$, $C_{12}(n,0)$	...
	1	$R_{12}(0,1)$, $C_{12}(0,1)$	$R_{12}(1,1)$, $C_{12}(1,1)$	$R_{12}(2,1)$, $C_{12}(2,1)$	...	$R_{12}(n,1)$, $C_{12}(n,1)$	...
	2	$R_{12}(0,2)$, $C_{12}(0,2)$	$R_{12}(1,2)$, $C_{12}(1,2)$	$R_{12}(2,2)$, $C_{12}(2,2)$	...	$R_{12}(n,2)$, $C_{12}(n,2)$	...
	...	...	...	...		...	...
	m	$R_{12}(0,n)$, $C_{12}(0,n)$	$R_{12}(1,m)$, $C_{12}(1,m)$	$R_{12}(2,m)$, $C_{12}(2,m)$	...	$R_{12}(n,m)$, $C_{12}(n,m)$	...
	...	...	...	...		...	...

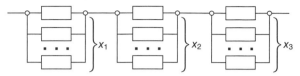

FIGURE 5.3 Block diagram of the system for the numerical example.

TABLE 5.5   Initial Dominating Sequences for Redundant Groups 1, 2, and 3

Group number		Number of redundant units in the group						
		0	1	2	3	4	5	...
1	R	0.7000	0.9100	0.9730	0.9919	0.9976	0.9992	...
	C	0	1	2	3	4	5	...
2	R	0.5000	0.7500	0.8750	0.9375	0.9688	0.9844	...
	C	0	3	6	8	12	15	...
3	R	0.5000	0.7500	0.8750	0.9375	0.9688	0.9844	...
	C	0	1	2	3	4	5	...

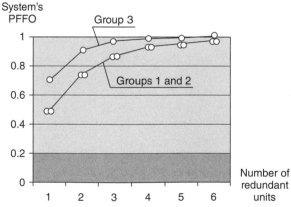

FIGURE 5.4 Concave shapes of R(C) functions for initial redundant groups.

TABLE 5.6 Dominating Sequence for the Composition of Groups 1 and 2

		x_2					
		0	1	2	3	4	5
x_1	0	0.350; 0 *1*	0.525; 3 *4*	0.613; 6	0.656; 9	0.678; 12	0.695; 15
	1	0.454; 1 *2*	0.683; 4 *5*	0.796; 7 *8*	0.853; 10	0.882; 13	0.896; 16
	2	0.487; 2 *3*	0.730; 5 *6*	0.851; 8 *9*	0.912 1 *12*	0.942; 14 *15*	0.958; 17
	3	0.496; 3 *7*	0.744; 6 *7*	0.868; 9 *10*	0.930; 12 *13*	0.961; 15 *16*	0.976; 18 *19*
	4	0.499; 4	0.748; 7	0.873; 10 *11*	0.935; 13 *14*	0.966; 16 *17*	0.981; 19 *20*
	5	0.500; 5	0.749; 8	0.874; 11	0.937; 14	0.968; 17 *18*	0.984; 20 *21*
		. . .	. . .	. . .	. . .	. . .	. . .

From Figure 5.5, one can see that the dominating sequence is not strictly concave, though there is some kind of concave envelope that, by the way, very often coincides with solutions obtained by the SDM.

On the basis of Table 5.6, one can construct Table 5.7, containing only dominating reliability-cost pairs.

Let us enumerate corresponding pairs of this dominating sequence with the number $x^{(1)}$. Now we have a system consisting of two redundant groups: group 3 (data are on the lower lines of Table 5.5) and the newly composed group (data are in Table 5.7).

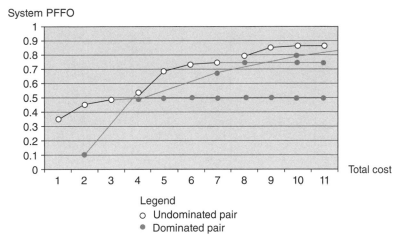

FIGURE 5.5 Graphical presentation of the data in the upper left corner of Table 5.6.

TABLE 5.7   Beginning of the Dominating Sequence in Table 5.6

Number	1	2	3	4	5	6	7	8	9	10	11	. . .
Domin. pair	0.350; 0	0.454; 1	0.487; 2	0.525; 3	0.683; 4	0.730; 5	0.744; 6	0.796; 7	0.851; 8	0.868; 9	0.873; 10	. . .
(x_1, x_2)	(0, 0)	(0, 1)	(0, 2)	(0, 3)	(1, 1)	(1, 2)	(1, 3)	(2, 1)	(2, 2)	(2, 3)	(2, 4)	. . .

On the basis of these data let us combine the final group for the considering system (Table 5.8 and Table 5.9).

5.2.3   Solving the Direct and Inverse Problems of Optimal Redundancy

Using Table 5.7, it is easy to get solutions for both direct and inverse problems of optimal redundancy (see Fig. 5.6). For instance, if one needs to find the best redundant unit allocation to satisfy the requirement $R \geq 0.8$, then from Table 5.7 we find that the solution for corresponding value ($R = 0.825$) is in cell ($x^{(1)} = 9$, $x_3 = 4$). The

TABLE 5.8 Final Data for Solving the Optimal Redundancy Problems for the Illustrative Example

	Number of dominating variant of pairs $\{x^{(1)} = (x_1, x_2)\}$									
	1	2	3	4	5	6	7	8	9	...
0	0.175; 0 *1*	0.227; 1	0.243; 2	0.262; 3	0.341; 4	0.365; 5	0.372; 6	0.398; 7	0.425; 8	...
1	0.263; 1 *2*	0.341; 2 *3*	0.365; 3	0.393; 4	0.512; 5 6	0.546; 6	0.556; 7	0.596; 8	0.638; 9	...
2	0.306; 2	0.398; 3 *4*	0.425; 4	0.458; 5	0.597; 6 *7*	0.639; 7	0.650; 8	0.696; 9	0.745; 10	...
3	0.330; 3	0.427; 4 5	0.456; 5	0.492; 6	0.640; 7 *8*	0.685; 8 *9*	0.705; 9	0.746; 10 *11*	0.798; 11 *12*	...
4	0.340; 4	0.441; 5	0.472; 6	0.510; 7	0.661; 8	0.707; 9 *10*	0.720; 10	0.772; 11	0.825; 12 *13*	...
5	0.343; 5	0.7448; 6	0.479; 7	0.516; 8	0.672; 9	0.719; 10	0.732; 11	0.784; 12	0.837; 13	...
...	...	...	...	...	...	...	...	...	...	...

x_3

TABLE 5.9 Final Dominating Sequence for the System

R	0.175	0.263	0.341	0.398	0.427	0.512	0.597	0.640	0.685	0.707	0.746	0.798	0.825	...
C	0	1	2	3	4	5	6	7	8	9	10	11	12	...

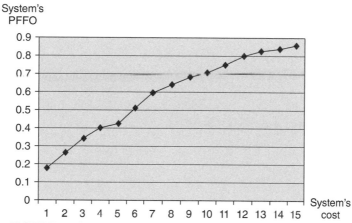

FIGURE 5.6 Final dominating sequence for the system.

cost of redundant units in this case is 12. In turn, for $x^{(1)} = 9$ one finds that this corresponds to $x_2 = 1$ and $x_2 = 2$.

Thus, the solution of the inverse problem is $(x_1 = 2, x_2 = 2, x_3 = 3)$.

If there is a limitation on the redundant units total cost equal to 10, then from Table 5.7 one finds that corresponding maximum value of probability of failure-free operation (PFFO) is 0.746. This solution corresponds to a cell $(x^{(1)} = 8, x_3 = 3)$. In turn, for $x^{(1)} = 8$ one finds that this corresponds to $x_1 = 1$ and $x_2 = 2$.

Thus, the solution of the inverse problem is $(x_1 = 1, x_2 = 2, x_3 = 3)$.

There are two ways of choosing redundant groups for composing a dominating sequence (see Fig. 5.7).

For computer solution, both types are equivalent. However if one needs to make calculations by hand, then the dichotomous way gives a substantial decrease in calculations.

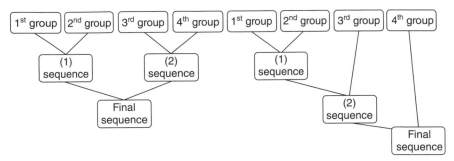

FIGURE 5.7 Two methods of choosing redundant groups for composing a dominating sequence.

CHRONOLOGICAL BIBLIOGRAPHY

Bellman, R. 1957. *Dynamic Programming*. Princeton University Press.

Bellman, R., and Dreyfus, S. E. 1962. *Applied Dynamic Programming*. Princeton University Press.

Kettelle, J. D., Jr.Jr. 1962. "Least-coast allocation of reliability investment." *Operations Research*, no. 2.

Li, J. 1966. "A bound dynamic programming for solving reliability redundancy optimization." *Microelectronics Reliability*, no. 10.

Ushakov, I. A. 1969. *Method of Solving Optimal Redundancy Problems under Constraints* (in Russian). Sovetskoe Radio.

Misra, K. B. 1971. "Dynamic programming formulation of the redundancy allocation problem." *International Journal of Mathematical Education in Science and Technology*, no. 3.

Yalaoui, A., Chatelet, E., and Chu, C. 2005. "A new dynamic programming method for reliability and redundancy allocation in a parallel-series system." *IEEE Transactions on Reliability*, no. 2.

CHAPTER 6

UNIVERSAL GENERATING FUNCTIONS

The Method of Universal Generating Functions (U-functions), introduced by Ushakov (1986), actually represents a generalization of Kettelle's Algorithm. This method suggests a transparent and convenient method of computerized solutions of various enumeration problems, in particular, the optimal redundancy problem.

6.1   GENERATING FUNCTION

First, let us refresh our memory concerning generating functions. This is a very convenient tool widely used in probability theory for finding joint distributions of several discrete random variables. Generating function is defined as:

$$\varphi(z) = \sum_{k \in G_k} p_k z^k, \qquad (6.1)$$

Optimal Resource Allocation: With Practical Statistical Applications and Theory,
First Edition. Igor A. Ushakov.
© 2013 John Wiley & Sons, Inc. Published 2013 by John Wiley & Sons, Inc.

where p_k is the probability that discrete random variable X takes value k and G_k is the distribution function domain. In optimal redundancy problems, in principle, $G_k = [0, \infty)$, though any practical task has its own limitations on the largest value of k.

Consider two non-negative discrete random values X_1 and X_2 with distributions

$$\begin{cases} P\{X^{(1)} = 1\} = p_1^{(1)}, \\ P\{X^{(1)} = 2\} = p_2^{(1)} \\ \qquad \cdots \\ P\{X^{(1)} = n_1\} = p_{n_1}^{(1)}, \end{cases}$$

and

$$\begin{cases} P\{X^{(2)} = 1\} = p_1^{(2)}, \\ P\{X^{(2)} = 2\} = p_2^{(2)}, \\ \qquad \cdots \\ P\{X^{(2)} = n_2\} = p_{n_2}^{(2)}, \end{cases}$$

correspondingly, where n_1 and n_2 are numbers of discrete values of each type. For finding probability of random variable $X = X^{(1)} + X^{(2)}$, one should enumerate all possible pairs of $X^{(1)}$ and $X^{(2)}$ that give in sum value k and add corresponding probabilities:

$X^{(1)} = 0$, $X^{(2)} = k$, with probability $p_0^{(1)} \cdot p_k^{(2)}$,
$X^{(1)} = 1$, $X^{(2)} = k - 1$, with probability $p_1^{(1)} \cdot p_{k-1}^{(2)}$,
$X^{(1)} = 2$, $X^{(2)} = k - 2$, with probability $p_2^{(1)} \cdot p_{k-2}^{(2)}$,

. . .

$X^{(1)} = k$, $X^{(2)} = 0$, with probability $p_k^{(1)} \cdot p_0^{(2)}$.

Thus the probability of interest is equal to

$$P\{X = k\} = \sum_{0 \leq j \leq k} p_j^{(1)} \cdot p_{k-j}^{(2)} = \sum_{0 \leq j \leq k} p_{k-j}^{(1)} \cdot p_j^{(2)}. \qquad (6.2)$$

One can see that there is a convolution transform. It is clear that the same result will be obtained if one takes a polynomial

$$\varphi(z) = \varphi^{(1)}(z) \cdot \varphi^{(2)}(z) \tag{6.3}$$

and, after combining like terms of expansions, finds the coefficient at z^k.

6.2 UNIVERSAL GF (U-FUNCTION)

One sees that algebraic argument "z" was introduced only for convenience: everybody knows that polynomial multiplication means products of coefficients and sums of powers. Such presentation helps one to obtain a distribution of the convolution of discrete random variables. However, what if random variables should be exposed to transformation different from convolution? For instance, if these random variables are arguments of some function?

Let us use habitual form of presentation, using symbol "$\otimes$" instead of "Π" just to underline that this is not an ordinary product of two GFs but a special transformation:

$$\varphi(z) = \varphi^{(1)}(z) \underset{f}{\otimes} \varphi^{(2)}(z) = \left(\sum_{1 \le i \le n_1} p_i^{(1)} z^{X_i^{(1)}} \right) \underset{f}{\otimes} \left(\sum_{1 \le j \le n_2} p_j^{(2)} z^{X_j^{(2)}} \right)$$
$$= \sum_{\substack{1 \le i \le n_1 \\ 1 \le j \le n_2}} p_i^{(1)} p_j^{(2)} z^{f(X_i^{(1)}, X_j^{(2)})}. \tag{6.4}$$

The subscript "f" in $\underset{f}{\otimes}$ means that some specific operation f will be undertaken over values X. It is clear that in the case of "pure" GF function is operation of summation.

In the general case, using the polynomial form of GF is inconvenient and even impossible. In order to move further, let us introduce some terms. We previously said that a system consists of units that are physical objects characterized by their parameters: reliability, cost, weight, and so on. So we can consider each unit as a multiplet of its parameter. Reliability of each unit can be improved by using redundancy or by replacing units with more effective units. In other words, on a design stage an engineer deals with a "string" of possible multiplets characterizing various variants of a considered unit.

Consider a series system of two units. Unit 1 and Unit 2 are characterized by strings:

$$S_1 = \{M_1^{(1)}, M_2^{(1)}, \dots, M_{n_1}^{(1)}\}$$

and

$$S_2 = \{M_1^{(2)}, M_2^{(2)}, \dots, M_{n_2}^{(2)}\}.$$

Each multiplet is a set of parameters $M_j^{(k)} = \{\alpha_{1j}^{(k)}, \alpha_{2j}^{(k)}, \dots, \alpha_N^{(k)}\}$ where N is the number of parameters in each multiplet.

"Interaction" of these two strings is an analogue of the Cartesian product whose members fill the cells of Table 6.1.

Interaction of multiplets consists of interaction of their similar parameters, for instance,

$$M_j^{(k)} \otimes M_i^{(h)} = \{(\alpha_{1j}^{(k)} \underset{f_1}{\otimes} \alpha_{1i}^{(h)}), (\alpha_{2j}^{(k)} \underset{f_2}{\otimes} \alpha_{2i}^{(h)}), \dots, (\alpha_{Nj}^{(k)} \underset{f_N}{\otimes} \alpha_{Ni}^{(h)})\}. \quad (6.5)$$

Operator $\otimes$, as well as each operator $\underset{f_s}{\otimes}$, in most natural practical cases possesses the commutativity property, that is,

$$\underset{f}{\otimes}(a, b) = \underset{f}{\otimes}(b, a), \quad (6.6)$$

and the associativity property, that is,

$$\underset{f}{\otimes}(a, b, c) = \underset{f}{\otimes}(a \underset{f}{\otimes}(b, c)) = \underset{f}{\otimes}((a \underset{f}{\otimes} b), c). \quad (6.7)$$

TABLE 6.1　"Interaction" of Two Strings

	$M_1^{(1)}$	$M_2^{(1)}$	$\cdots$	$M_{n_1}^{(1)}$
$M_1^{(2)}$	$M_1^{(1)} \otimes M_1^{(2)}$	$M_2^{(1)} \otimes M_1^{(2)}$	$\cdots$	$M_{n_1}^{(1)} \otimes M_1^{(2)}$
$M_2^{(2)}$	$M_1^{(1)} \otimes M_2^{(2)}$	$M_2^{(1)} \otimes M_2^{(2)}$	$\cdots$	$M_{n_1}^{(1)} \otimes M_2^{(2)}$
$\cdots$	$\cdots$	$\cdots$	$\cdots$	$\cdots$
$M_{n_2}^{(2)}$	$M_1^{(1)} \otimes M_{n_2}^{(2)}$	$M_2^{(1)} \otimes M_{n_2}^{(2)}$	$\cdots$	$M_{n_1}^{(1)} \otimes M_{n_2}^{(2)}$

Of course, operator $\underset{f_s}{\otimes}$ depends on the physical nature of parameter α_s and the type of structure, that is, series or parallel (see Table 6.2).

In the problem of optimal redundancy, one deals with triplet of type "Probability–Cost–Number of units" for each redundant

TABLE 6.2   "Interaction" of Various Parameters for Series and Parallel Connections

Type of parameter	Type of structure	Result of interaction
(A) α is unit's PFFO (probability of failure-free operation)	Series	$\alpha_{Aj}^{(k)} \underset{f}{\otimes} \alpha_{Ai}^{(h)} = \alpha_{Aj}^{(k)} \times \alpha_{Ai}^{(h)}$
	Parallel	$\alpha_{Aj}^{(k)} \underset{f}{\otimes} \alpha_{Ai}^{(h)} = 1 - (1 - \alpha_{Aj}^{(k)}) \times (1 - \alpha_{Ai}^{(h)})$
(B) α is number of units in parallel	Series	$\alpha_{Bj}^{(k)} \underset{f}{\otimes} \alpha_{Bi}^{(h)} = (\alpha_{Bj}^{(k)}; B_{Ri}^{(h)})$
	Parallel	$\alpha_{Bj}^{(k)} \underset{f}{\otimes} \alpha_{Bi}^{(h)} = (\alpha_{Bj}^{(k)}; B_{Ri}^{(h)})$
(C) α is unit's cost (weight)	Series	$\alpha_{Aj}^{(k)} \underset{f}{\otimes} \alpha_{Ai}^{(h)} = \alpha_{Aj}^{(k)} + \alpha_{Ai}^{(h)}$
	Parallel	$\alpha_{Aj}^{(k)} \underset{f}{\otimes} \alpha_{Ai}^{(h)} = \alpha_{Aj}^{(k)} + \alpha_{Ai}^{(h)}$
(D) α is unit's ohmic resistance	Series	$\alpha_{Aj}^{(k)} \underset{f}{\otimes} \alpha_{Ai}^{(h)} = \alpha_{Aj}^{(k)} + \alpha_{Ai}^{(h)}$
	Parallel	$\alpha_{Aj}^{(k)} \underset{f}{\otimes} \alpha_{Ai}^{(h)} = [(\alpha_{Aj}^{(k)})^{-1} + (\alpha_{Ai}^{(h)})^{-1}]^{-1}$
(E) α is unit's capacitance	Series	$\alpha_{Aj}^{(k)} \underset{f}{\otimes} \alpha_{Ai}^{(h)} = [(\alpha_{Aj}^{(k)})^{-1} + (\alpha_{Ai}^{(h)})^{-1}]^{-1}$
	Parallel	$\alpha_{Aj}^{(k)} \underset{f}{\otimes} \alpha_{Ai}^{(h)} = \alpha_{Aj}^{(k)} + \alpha_{Ai}^{(h)}$
(F) α is pipeline unit's capacitance	Series	$\alpha_{Aj}^{(k)} \underset{f}{\otimes} \alpha_{Ai}^{(h)} = \min\{\alpha_{Aj}^{(k)}, \alpha_{Ai}^{(h)}\}$
	Parallel	$\alpha_{Aj}^{(k)} \underset{f}{\otimes} \alpha_{Ai}^{(h)} = \alpha_{Aj}^{(k)} + \alpha_{Ai}^{(h)}$
(G) α is unit's random time to failure	Series	$\alpha_{Aj}^{(k)} \underset{f}{\otimes} \alpha_{Ai}^{(h)} = \min\{\alpha_{Aj}^{(k)}, \alpha_{Ai}^{(h)}\}$
	Parallel	$\alpha_{Aj}^{(k)} \underset{f}{\otimes} \alpha_{Ai}^{(h)} = \max\{\alpha_{Aj}^{(k)}, \alpha_{Ai}^{(h)}\}$

group: $M_j = \{\alpha_{1j}, \alpha_{2j}, \alpha_{3j}\}$. If there is a system of n series subsystems (single elements or redundant groups), one has to use a procedure almost completely coincided with the procedure of compiling the dominating sequences at Kettelle's algorithm. In other words, the problem reduces to constructing a single "equivalent unit" that possesses the entire system's properties. There are two possible ways of "convolving" the system into a single "equivalent unit": dichotomous and sequential. We will demonstrate these two possible procedures in an example of a series system of four subsystems (see descriptions in Fig. 6.1 and Fig. 6.2).

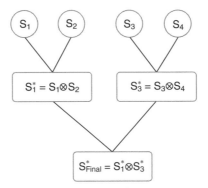

FIGURE 6.1 Dichotomous scheme of compiling the equivalent unit.

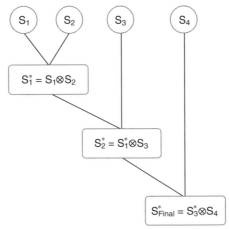

FIGURE 6.2 Dichotomous scheme of compiling the equivalent unit.

TABLE 6.3   System Unit Parameters

	Unit 1	Unit 2	Unit 3	Unit 4
PFFO	0.6	0.6	0.7	0.7
Cost	3	1.5	2	1.2

Example 6.1

Consider a series system of four units with parameters as given in Table 6.3. Assume that "hot" redundancy of each unit is used for the system reliability improvement.

Let us solve two problems of optimal redundancy:

(a) Find the optimal allocation of redundant units to reach required PFFO level of 0.95.

(b) Find the optimal allocation of redundant units to reach maximum possible PFFO (probability of failure-free operation) level under the condition that the total cost of the system does not exceed 30 cost units.

In this case, each unit is characterized by the following strings of triplets (cost, PFFO, number):

$$S_1 = [\{3; 0.6; 1\}, \ldots, \{12; 0.9744; 4\}, \{15; 0.9898; 5\}, \{18; 0.9959; 6\},$$
$$\{21; 0.9984; 7\}, \{24; 0.9993; 8\}, \{27; 0.9997; 9\}, \ldots]$$

$$S_2 = [\{1.5; 0.6; 1\}, \ldots, \{6; 0.9744; 4\}, \{7.5; 0.9898; 5\}, \{9; 0.9959; 6\},$$
$$\{10.5; 0.9984; 7\}, \{12; 0.9993; 8\}, \{13.5; 0.9997; 9\}, \ldots]$$

$$S_3 = [\{2; 0.7; 1\}, \ldots, \{6; 0.9730; 3\}, \{8; 0.9919; 4\}, \{10; 0.9976; 5\},$$
$$\{12; 0.9993; 6\}, \{14; 0.9998; 7\}, \{16; 0.9999; 8\}, \ldots]$$

$$S_4 = [\{1.2; 0.7; 1\}, \ldots, \{3.6; 0.9730; 3\}, \{4.8; 0.9919; 4\}, \{6; 0.9976;$$
$$5\}, \{7.2; 0.9993; 6\}, \{8.4; 0.9998; 7\}, \{9.6; 0.9999; 8\}, \ldots].$$

Let us apply the dichotomous scheme of compiling equivalent units, and, first, consider the subsystem consisting of Unit 1 and

Unit 2 (Table 6.4). We omit all intermediate calculations performed with the help of a simple Excel program.

In Table 6.4, as well as in Tables 6.5 and 6.6, shaded entries indicate those triplets which are dominated. In the same manner, we construct S_2^* (Table 6.5).

TABLE 6.4   Triplets Belonging to S_1^*

Cost	PFFO	$X = (x_1, x_2)$
...	...	...
18	0.9495	(4,4)
19.5	0.9644	(4,5)
21	0.9704	(4,6)
21	0.9644	(5,4)
22.5	0.9796	(5,5)
22.5	0.9728	(4,7)
24	0.9857	(5,6)
24	0.9704	(6,4)
24	0.9738	(4,8)
25.5	0.9881	(5,7)
25.5	0.9857	(6,5)
25.5	0.9741	(4,9)
27	0.9728	(7,4)
27	0.9918	(6,6)
27	0.9891	(5,8)
28.5	0.9881	(7,5)
28.5	0.9943	(6,7)
28.5	0.9895	(5,9)
30	0.9738	(8,4)
30	0.9943	(7,6)
30	0.9953	(6,8)
...	...	...

TABLE 6.5   Triplets Belonging to S_2^*

Cost	PFFO	$X = (x_3, x_4)$
...	...	...
9.6	0.9467	(3, 3)
10.8	0.9651	(3, 4)
11.6	0.9651	(4, 3)
12	0.9706	(3, 5)
12.8	0.9839	(4, 4)
13.2	0.9723	(3, 6)
13.6	0.9706	(5, 3)
14	0.9895	(4, 5)
14.4	0.9728	(3, 7)
14.8	0.9895	(5, 4)
15.2	0.9912	(4, 6)
15.6	0.9723	(6, 3)
15.6	0.9729	(3, 8)
16	0.9951	(5, 5)
...	...	...

Now on the basis of S_1^* and S_2^*, one constructs the final string for the equivalent unit. The result is given in the Table 6.6. One can notice that the final string in this particular case completely coincides with the final dominating sequence obtained by the Kettelle's Algorithm: the only difference is that we kept "the track of solving process" and have the resulting solution immediately from the table. (Frankly speaking, Kettele's Algorithm could be easily modified to get the same property of the final solution.)

Solutions of the problems above can be easily found from Table 6.6. The first time PFFO exceeds a level of 0.95 is when $X = (4,5,5,4)$ and the corresponding system cost is 33.5 cost units. The second task has solution $X = (4,4,5,3)$ with the cost equal to exactly 30 cost units. PFFO for this case is equal to 0.9216.

TABLE 6.6   Resulting String of Triplets for the Equivalent Unit

Cost	PFFO	$X = (x_3, x_4, x_3, x_4)$
...	...	...
27.6	0.8989	(4, 4, 2, 3)
28.8	0.9164	(4, 4, 4, 3)
29.1	0.9164	(4, 5, 2, 3)
30.0	0.9216	(4, 4, 5, 3)
30.3	0.9307	(4, 5, 4, 3)
30.6	0.9187	(3, 6, 2, 3)
30.8	0.9342	(4, 4, 4, 8)
31.5	0.9360	(4, 5, 5, 3)
31.8	0.9365	(3, 6, 4, 3)
32.0	0.9395	(4, 4, 5, 4)
32.1	0.9274	(5, 5, 2, 3)
32.3	0.9489	(4, 5, 4, 8)
33.0	0.9419	(3, 6, 5, 3)
33.2	0.9411	(4, 4, 6, 4)
33.3	0.9454	(5, 5, 4, 3)
33.5	0.9543	(4, 5, 5, 4)
33.6	0.9389	(5, 6, 2, 3)
33.8	0.9548	(3, 6, 4, 8)
...	...	...

As the conclusion of this chapter, let us notice that the U-function method is very constructive not only for solving optimal redundancy problems, but also for a number of other problems, particularly those associated with multi-state systems analysis.

CHRONOLOGICAL BIBLIOGRAPHY

Ushakov, I. A. 1986. "A Universal Generating Function" (in Russian). *Engineering Cybernetics*, no. 5.

Ushakov, I. A. 1987. "Universal generating function." *Soviet Journal of Computer and System Science*, no. 3.

Ushakov, I. A. 1987. "Optimal standby problem and a universal generating function." *Soviet Journal of Computer and System Science*, no. 4.

Ushakov, I. A. 1987. "Solution of multi-criteria discrete optimization problems using a universal generating function." *Soviet Journal of Computer and System Sciences*, no. 5.

Ushakov, I. A. 1988. "Solving of optimal redundancy problem by means of a generalized generating function." *Elektronische Informationsverarbeitung und Kybernetik*, no. 4–5.

Ushakov, I. A. 2000. "The method of generating sequences." *European Journal of Operational Research*, vol. 125, no. 2.

Levitin, G. 2005. *The Universal Generating Function in Reliability Analysis and Optimisation*. Springer-Verlag.

Chakravarty, S., and Ushakov, I. A. 2008. "Object Oriented Commonalities in Universal Generating Function for Reliability and in C++." *Reliability and Risk Analysis: Theory and Applications*, no. 10.

CHAPTER 7

GENETIC ALGORITHMS

7.1 INTRODUCTION

Computer simulations of evolution started in the mid-1950s, when a Norwegian-Italian mathematician, Nils Aall Barricelli (1912–1993), who had been working at the Institute for Advanced Study in Princeton, published his first paper on the subject. This was followed by a series of works that were published in the 1960s–1970s.

Genetic algorithms (GAs) became especially popular through the work of John Holland (see Box) in the early 1970s, and particularly because of his book *Adaptation in Natural and Artificial Systems* (1975).

Optimal Resource Allocation: With Practical Statistical Applications and Theory, First Edition. Igor A. Ushakov.

John Henry "Dutchy" Holland (Born in 1929)

John Holland was an American scientist, Professor of Psychology, and Professor of Electrical Engineering and Computer Science at the University of Michigan, Ann Arbor. He is a pioneer in complex system and nonlinear science and is known as the father of genetic algorithms. In 1975 he wrote his book on genetic algorithms, *Adaptation in Natural and Artificial Systems*.

Holland wrote: "A Genetic Algorithm is a method of problem analysis based on Darwin's theory of natural selection. It starts with an initial population of individual nodes, each with randomly generated characteristics. Each is evaluated by some method to see which ones are more successful. These successful ones are then merged into one 'child' that has a combination of traits of the parents' characteristics."

In recent years, many studies on reliability optimization use a universal optimization approach based on metaheuristics. Genetic algorithms are considered as a particular class of evolutionary algorithms that use techniques inspired by Darwin's evolution theory in biology and include such components as inheritance, mutation, selection, and crossover (recombination).

These metaheuristics hardly depend on the specific nature of the problem that is being solved and, therefore, can be easily applied to solve a wide range of optimization problems. The metaheuristics are based on artificial reasoning rather than on classical mathematical programming. An important advantage of these methods is that they do not require any prior information and are based on collection of current data obtained during the randomized search process. These data are substantially used for directing the search.

Genetic algorithms are implemented as a computer simulation in which a population of abstract items (called "chromosomes" or "the genotype of the genome") represent "candidate solutions" (called individuals, creatures, or phenotypes) systematically lead toward better solutions.

GAs have the following advantages in comparison with traditional methods:

(1) They can be easily implemented and adapted.
(2) They usually converge rapidly on solutions of good quality.
(3) They can easily handle constrained optimization problems.

A GA requires strong definition of two things:

(1) A genetic representation of the solution domain
(2) A fitness function to evaluate the solution domain.

The fitness function is defined over the genetic representation and measures the quality of the presented solution. The fitness function always depends on the problem's nature. In some problems, it is impossible to define the fitness expression, so one needs to use interactive procedures based on expert's opinion.

As soon as we have the genetic representation and the properly defined fitness function, a GA proceeds to initialize a population of solutions randomly.

A genetic algorithm includes the following main phases.

7.1.1 Initialization

A number of individual solutions are generated at random to form an initial population. The population size depends on the nature of the problem, but typically contains hundreds or thousands of possible solutions. The population is generated to be able to cover the entire range of possible solutions (the search space). Occasionally, some solutions may be "seeded" in areas where actual optimal solution is located.

This initial population of solutions is undertaken to improve the procedure through repetitive application of selection, reproduction, mutation, and crossover operators.

7.1.2 Selection

Obtained individual solutions are selected through a special process using a fitness function that allows ordering the solutions by specified quality measure. These selection methods rate the fitness of each solution and preferentially select the best solutions.

7.1.3 Reproduction

The next step is generating the next generation of solutions from those selected through genetic operators: crossover (recombination) and mutation.

Each new solution is produced by a pair of "parent" solutions selected for "breeding." By producing a "child" solution using the above methods of crossover and mutation, a new solution is created that typically shares many of the characteristics of its "parents." New parents are selected for each child, and the process continues until a new population of solutions of appropriate size is generated.

7.1.4 Termination

The reproduction process is repeated until a termination condition has been reached. Common terminating conditions are:

(1) The predetermined number of produced generations has been reached, or

(2) A satisfactory fitness level has been reached for the population.

7.2 STRUCTURE OF STEADY-STATE GENETIC ALGORITHMS

The steady-state GA (see Fig. 7.1) proceeds as follows: an initial population of solutions is generated randomly or heuristically.[1] Within this population, new solutions are obtained during the genetic cycle by using the crossover operator. This operator produces an offspring from a randomly selected pair of parent solutions that are selected with a probability proportional to their relative fitness. The newly obtained offspring undergoes mutation with the probability p_{mut}.

Each new solution is decoded and its fitness function value is estimated. These values are used for a selection procedure that determines what is better: the newly obtained solution or the worst

---

[1] Material for this section is presented by G. Levitin.

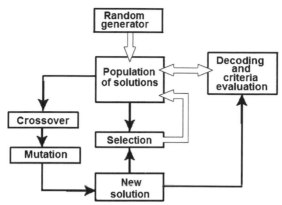

FIGURE 7.1 Structure of a steady-state GA.

solution in the population. The better solution joins the population, while the current one is discarded. If the solution population contains a pair of equivalent items, then either of them is eliminated and the population size decreases.

The stopping rule is when the number of new solutions reaches some level N_{rep}, or when the number of remaining solutions in the population after excluding reaches a specified level. After this, the new genetic cycle begins: a new population of randomly constructed solutions are generated and the process continues. The whole optimization process terminates when its specified termination condition is satisfied. This condition can be specified in the same way as in a generational GA.

The steady-state GA can be presented in the following pseudo-code format.

```
begin STEADY STATE GA
   Initialize population Π
   Evaluate population Π   {compute fitness values}
   while GA termination criterion is not satisfied do
         {GENETIC CYCLE}
         while genetic cycle termination criterion is
not satisfied do
```

```
                    Select at random Parent Solu-
tions S₁, S₂ from Π
                    Crossover: (S₁, S₂) → S₀
{offspring}
                    Mutate offspring S₀ → S*₀
with probability p_mut
                    Evaluate S*₀
                    Replace S_w {the worst solu-
tion in Π  with S*₀ } if S*₀ is
                    better than S_w
                    Eliminate identical solutions
in Π
        end while
        Replenish Π  with new randomly generated
solutions
  end while
end GA
```

7.3 RELATED TECHNIQUES

Here is a list (in alphabetic order) of a number of techniques "genetically" close to genetic algorithm.[2]

Ant colony optimization (ACO) uses many ants (or agents) to traverse the solution space and find locally productive areas.

Bacteriologic algorithms (BA) inspired by evolutionary ecology and, more particularly, bacteriologic adaptation.

Cross-entropy method (CE) generates candidates solutions via a parameterized probability distribution.

Deluge algorithm (GD) is a generic algorithm similar in many ways to the hill-climbing and simulated annealing algorithms.

---

[2]This section is based partly on http://en.wikipedia.org/wiki/Genetic_algorithm.

Evolution strategies (ES) evolve individuals by means of mutation and intermediate and discrete recombination.

Evolutionary programming (EP) involves populations of solutions with primarily mutation and selection and arbitrary representations.

Extremal optimization (EO) evolves a single solution and makes local modifications to the worst components.

Genetic programming (GP) is a technique in which computer programs, rather than function parameters, are optimized.

Grouping genetic algorithm (GGA) is an evolution of the GA where the focus is shifted from individual items, as in classical GAs, to groups or subsets of items.

Harmony search (HS) is an algorithm mimicking musicians' behaviors in improvisation processes.

Immune optimization algorithm (IOA) is based on both the concept of Pareto optimality and simple interactive metaphors between antibody populations and multiple antigens.

Interactive evolutionary algorithms (IEA) are evolutionary algorithms that use human evaluation when it is hard to design a computational fitness function.

Mimetic algorithm (MA) is a relatively new evolutionary method where local search is applied during the evolutionary cycle.

Particle swarm optimization (PSO) is an algorithm to find a solution to an optimization problem in a search space or model and predict social behavior in the presence of objectives.

Simulated annealing (SA) is a related global optimization technique that traverses the search space by testing random mutations on an individual solution.

Taboo search (TS) is similar to simulated annealing in that both traverse the solution space by testing mutations of an individual solution.

CHRONOLOGICAL BIBLIOGRAPHY

Coit, D. W., and Smith, A. E. 1966. "Reliability optimization of series-parallel systems using a genetic algorithm." *IEEE Transactions on Reliability*, no. 45

Coit, D. W., Smith, A. E. and Tate, D. M. 1966. "Adaptive penalty methods for genetic optimization of constrained combinatorial problems." *INFORMS Journal of Computing*, no. 8.

Kumar, A., Pathak, R., Gupta, and Parsaei, H. 1995. "A genetic algorithm for distributed system topology design." *Computers and Industrial Engineering*, no 3.

Kumar, A., Pathak, R., and Gupta, Y. 1995. "Genetic algorithm-based reliability optimization for computer network expansion." *IEEE Transactions on Reliability*, no. 1.

Painton, L., and Campbell, J. 1995. "Genetic algorithm in optimization of system reliability." *IEEE Transactions on Reliability*, no. 2.

Coit, D. W., and Smith, A. E. 1996. "Penalty guided genetic search for reliability design optimization." *Computers and Industrial Engineering*, no. 30.

Coit, D. W., and Smith, A. E. 1996. "Reliability optimization of series-parallel systems using genetic algorithm." *IEEE Transactions on Reliability*, no. 2.

Coit, D. W., and Smith, A. E. 1996. "Solving the redundancy allocation problem using a combined neural network/genetic algorithm approach." *Computers and OR*, no. 6.

Dengiz, B., Altiparmak, F., and Smith, A. 1997. "Efficient optimization of all-terminal reliable networks using an evolutionary approach." *IEEE Transactions on Reliability*, vol. 46, no. 1.

Dengiz, B., Altiparmak, F., and Smith, A. 1997. "Local search genetic algorithm for optimal design of reliable networks." *IEEE Transactions on Evolutionary Computation*, no. 3.

Gen, M., and Cheng, R. 1997. *Genetic Algorithms and Engineering Design*. John Wiley & Sons.

Gen, M., and Kim, J. 1997. "GA-based reliability design: state-of-the-art survey." *Computers and Industrial Engineering*, no 1/2.

Ramachandran, V., Sivakumar, V., and Sathiyanarayanan, K. 1997. "Genetics based redundancy optimization." *Microelectronics and Reliability*, no 4.

Shelokar, P. S., Jayaraman, V. K., and Kulkarni, B. D. 1997. "Ant algorithm for single and multi-objective reliability optimization problems." *Quality and Reliability Engineering International*, no. 6

Coit, D. W., and Smith, A.E. 1998. "Redundancy allocation to maximize a lower percentile of the system time-to-failure distribution." *IEEE Transactions on Reliability*, no. 1.

Hsieh, Y., Chen, T., and Bricker, D. 1998. "Genetic algorithms for reliability design problems." *Microelectronics and Reliability*, no. 10.

Taguchi, T., Yokota, T., and Gen, M. 1998. "Reliability optimal design problem with interval coefficients using hybrid genetic algorithms." *Computers and Industrial Engineering*, no. 1/2.

Yang, J., Hwang, M., Sung, T., and Jin, Y. 1999. "Application of genetic algorithm for reliability allocation in nuclear power plant." *Reliability Engineering and System Safety*, no. 3.

Levitin, G., Kalyuzhny, A., Shenkman, A., and Chertkov, M. 2000. "Optimal capacitor allocation in distribution systems using a genetic algorithm and a fast energy loss computation technique." *IEEE Transactions on Power Delivery*, no. 2.

Levitin G., Dai,Y., Xie, M., and Poh, K. L. 2003. "Optimizing survivability of multistate systems with multi-level protection by multi-processor genetic algorithm." *Reliability Engineering and System Safety*, no. 2.

Liang, Y., and Smith, A. 2004. "An ant colony optimization algorithm for the redundancy allocation problem." *IEEE Transactions on Reliability*, no. 3.

Coit, D.W., and Baheranwala, F. 2005. "Solution of stochastic multi-objective system reliability design problems using genetic algorithms." In *Advances in Safety and Reliability*, K. Kolowrocki, ed. Taylor & Francis Group.

Levitin, G. 2005. "Genetic algorithms in reliability engineering." *Reliability Engineering and System Safety*, no. 2.

Marseguerra, M., Zio, E., and Podofillini, L. 2005. "Multi-objective spare part allocation by means of genetic algorithms and Monte Carlo simulation." *Reliability Engineering and System Safety*, no. 87.

Gupta, R., and Agarwal, M. 2006. "Penalty guided genetic search for redundancy optimization in multi-state series-parallel power system." *Journal of Combinatorial Optimization*, no. 3.

Levitin, G., ed. 2006. *Computational Intelligence in Reliability Engineering: Evolutionary Techniques in Reliability Analysis and Optimization*. Series: Studies in Computational Intelligence, vol. 39. Springer-Verlag.

Levitin, G., ed. 2006. *Computational Intelligence in Reliability Engineering: New Metaheuristics, Neural and Fuzzy Techniques in Reliability*. Series: Studies in Computational Intelligence, vol. 40. Springer-Verlag.

Levitin, G. 2006. "Genetic algorithms in reliability engineering." *Reliability Engineering and System Safety*, vol. 91, no. 9.

Konak, A., Coit, D., and Smith, A. 2006. "Multi-objective optimization using genetic algorithms: a tutorial." *Reliability Engineering and System Safety*, vol. 91, no. 9.

Parkinson, D. 2006. "Robust design employing a genetic algorithm." *Quality and Reliability Engineering International*, no. 3.

Taboada, H. A., Espiritu, J. F., and Coit, D. W. 2008. "A multi-objective multi-state genetic algorithm for system reliability optimization design problems." *IEEE Transactions on Reliability*, no. 1.

CHAPTER 8

MONTE CARLO SIMULATION

8.1 INTRODUCTORY REMARKS

Very often a reliability goal function cannot be expressed in a convenient analytical form and makes even calculation of reliability decrements practically impossible. For instance, such situations arise when system units are mutually dependent or their reliability simultaneously depends on some common environmental factors (temperature, mechanical impacts, etc.). In these cases, the Monte Carlo simulation is usually used for reliability indices calculation. However, the problem arises: how can one use the Monte Carlo simulation for optimization?

Roughly speaking, the idea is in observing the process of the spare unit expenditure (replacement of failed units) until specified restrictions allow one to do so. This may be a simulation process or an observation of the real deployment of the system. After the stopping moment, we start another realization of simulation process

Optimal Resource Allocation: With Practical Statistical Applications and Theory,
First Edition. Igor A. Ushakov.
© 2013 John Wiley & Sons, Inc. Published 2013 by John Wiley & Sons, Inc.

or observation of the real data. When the appropriate statistical data are collected, the process of finding the optimal solution starts.

Avoiding a formal description of the algorithm, let us demonstrate it on numerical examples which will make the idea of the method and its specific technique clearer.

8.2 FORMULATION OF OPTIMAL REDUNDANCY PROBLEMS IN STATISTICAL TERMS

Standard methods do not give a solution if the goal function is the mean time to failure:

$$T(x) = \int_0^\infty \prod_{1 \le i \le n} R_i(t \mid x_i)dt \qquad (8.1)$$

or if units are dependent, for instance, via a vector of some external factors g (temperature, humidity, vibration, etc.):

$$R(x) = \int_{g \in G} R(x_1, x_2, \dots x_n \mid g)dF(g), \qquad (8.2)$$

where $F(g)$ is the distribution function of some external parameter g and G is its domain.

8.3 ALGORITHM FOR TRAJECTORY GENERATION

For solution of the formulated problems, we need to have data obtained from real experiment (or system deployment) or from Monte Carlo simulation of the system model. Though this procedure is routine, we will briefly describe it for the presentation closeness. The procedure is as follows.

Consider a series system of n units. (For simplicity of explanation of the algorithm, we will assume that the units are independent. However, everything described below can be easily extended to the general case: it will effect only a mechanism of random sequence generation.)

Let us consider the process of spare unit expenditure as the process of changing the system states and the total cost of spare units at sequential replacement moments. After failure each unit is immediately replaced with a spare one. Let $t_k^{(j)}$ be the moment of the kth replacement during the jth Monte Carlo experiment. The number of spare units of type i spent at moment $t_k^{(j)}$ is denoted $x_{ik}^{(j)}$.

An initial state at $t_0^{(j)} = 0$ is:

$$x_{i0}^{(j)} \text{ for all } i, 1, 2, \ldots 1, \text{ and } j, j = 1, 2, \ldots, N.$$

The total cost of spare units at the initial moment is $C_0 = 0$. (Sometimes it might be reasonable to consider the initial cost of the system with no spare units as C_0, that is, $C_0 = \sum_{1 \leq i \leq n} k_i c_i$ where k_i is the number of units of type i within equipment before reliability improvement.)

Consider the step-by-step procedure of generating trajectories $\psi^{(s)}$, $s = 1, 2, \ldots$. We begin with $\psi^{(1)}$ but the corresponding superscript, (1), will be omitted for the sake of convenience.

Step 1. Generate random time to failure (TTF) for each unit, and define $b_{i1} = \xi_i$, that is b_{i1} is the moment of the earliest failure (and instantaneous replacement) of the ith unit. The current moment (for every unit i) at the beginning of any trajectory $\psi^{(s)}$ is $b_{i0} = 0$.

Step 2. Determine the moment of the occurrence of the first event (first replacement) within the first realization $\psi^{(1)}$ as $t_1 = \min_{1 \leq i \leq n} b_{i1}$.

Step 3. Assign to the corresponding unit (for which the moment of failure is the earliest one) a specific number $i = i_1$.

Step 4. Put into the spare units counter a new value $x_{i_1 1} = x_{i_1 0} + 1$.

Step 5. Rename remaining x_{i0} as follows: $x_{i0} = x_{i1}$ for all $i \neq i_1$.

Step 6. Calculate a new value of the system cost $C_1 = C_0 + C_{i_1}$.

Step 7. Generate the next random TTF for unit i_1, ξ_{i_1}.

Step 8. Calculate the next event occurring due to unit i_1:
$$b_{i_1 2} = t_1 + \xi_{i_1}.$$

Step 9. Rename the remaining values $b_{i1} = b_{i2}$ for all $i \neq i_1$.

Thus completes the first cycle. GOTO step 2, that is, find $t_2 = \min_{1 \leq i \leq n} b_{i2}$, and so on, until stopping the first realization.

The type of problem to be solved determines the stopping rule of each realization.

Stopping rule for the inverse problem of optimal redundancy: The process is stopped at the moment t_N when the total cost of spare units exceeds the permitted C^0.

Stopping rule for the direct problem of optimal redundancy: The simulation process for each realization stops at the moment $t_{M-1} < t^* \leq t_M$ where t^* equals the required operational time t_0 (if the reliability index is the probability of failure free operation) or t^* is the required system's mean time to failure (MTTF).

After the termination of generating the first trajectory, $\psi^{(1)}$, we start to generate $\psi^{(2)}$ by the same rules. The number of needed realizations, N, is determined by the required accuracy of statistical estimates.

Thus, each trajectory j represents a set of the following data:

$$\{t_1^{(j)}; X_1^{(j)}; C(X_1^{(j)})\}$$
$$\{t_2^{(j)}; X_2^{(j)}; C(X_2^{(j)})\}$$
$$\cdots$$
$$\{t_M^{(j)}; X_M^{(j)}; C(X_M^{(j)})\}$$

where $X_s^{(j)}$ is the set of spare units at moment $t_s^{(j)}$, that is, $X_s^{(j)} = \{x_{1s}^{(j)}, x_{2s}^{(j)}, \ldots, x_{ns}^{(j)}\}$.

After the description of the Monte Carlo simulating process, let us consider the optimization problems themselves. We can make an important remark: previously all problems were formulated in probabilistic terms, but dealing with statistical (empirical) functions has to be specific. The following problems are reformulated in an appropriate way.

8.4   DESCRIPTION OF THE IDEA OF THE SOLUTION

Assume that we need to supply some system with spare units for a specified period of time. We have no prior knowledge of units' reliability but we have an opportunity to observe a real process (or simulation) of failure occurrence.

Consider the direct problem of optimal redundancy. What shall we do in this case? We observe the process of spare unit expenditure during time t^*. This process can be described as a random path—call it trajectory—in a discrete n-dimensional space, X. An illustration of such a process in a two-dimensional case is presented in Figure 8.1.

Let us observe N such trajectories, $j = 1, 2, \ldots, N$, in an n-dimensional space where n is the number of unit types. Each

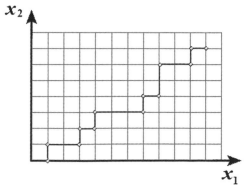

FIGURE 8.1   Example of a two-dimensional trajectory.

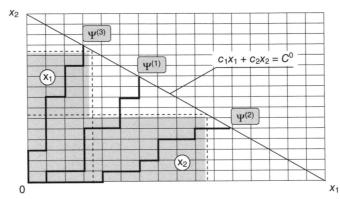

FIGURE 8.2 Example of two-dimensional trajectory inside hyperplane $C(x_1, x_2) = C^0$.

realization is stopped when the total cost of spare units exceeds the permitted amount, that is, each trajectory reaches or even penetrates a hyperplane:

$$\Gamma = \{C(X) = C^0\} \tag{8.3}$$

determined by the restriction on the total system cost (an example for the two-dimensional case is given in Fig. 8.2).

After this, in the same n-dimensional space, we construct the hypercube x_r, $r = 1, 2, \ldots$, such that each of its vertices is lying under the hyperplane. In Figure 8.2 there are two such hypercubes, though there could be many of them: actually, in this case all pairs (x_1, x_2) that belong to hyper plane $C(x_1, x_2) = C^0$ could be vertices of such a hypercube.

Denote the maximum time that trajectory $\psi^{(j)}$ is spending within the hypercube χ_r by $\tau_r^{(j)} = \tau(\psi^{(j)}, \tau_r)$. Introduce an indicator:

$$\delta_r^{(j)} = \begin{cases} 1 & \text{if } \tau_r^{(j)} \geq t_0 \\ 0 & \text{otherwise.} \end{cases} \tag{8.4}$$

Among all hypercubes above we choose the hypercube χ_r that maximizes the frequency of failure-free operation during required interval t_0 under the cost restrictions:

$$\max_{\chi_r}\left\{\frac{1}{N}\sum_{1\leq j\leq N}\delta_r^{(j)}\left|\sum_{\substack{1\leq i\leq n\\ x_i\in\chi_r}}x_ic_i\leq C^0\right|\right\}, \tag{8.5}$$

where C^0 is the admissible redundant group cost.

Maximization of the system average time to failure is reached by the hypercube χ_r' that corresponds to the solution of the following problem of the conditional optimization:

$$\max_{\chi_r'}\left\{\frac{1}{N}\sum_{1\leq j\leq N}\tau_r^{(j)}\left|\sum_{\substack{1\leq i\leq n\\ x_i\in\chi_r'}}x_ic_i\leq C^0\right|\right\}. \tag{8.6}$$

Now consider the direct problem of optimal redundancy. In this case, the required time of the system failure-free operation equals t_0. We observe the process of spare units expenditure N times until the system failure-free time exceeds t_0 and record trajectories $\psi^{(j)}$, $j = 1, 2, \ldots, N$, in an n-dimensional space. Afterwards we construct such hypercubes χ_r, $r = 1, 2, \ldots$, in the same n-dimensional space that includes (covers) $R^0 \times 100\%$ of all trajectories where R^0 is the specified level of the reliability index. Among all hypercubes described above, we choose the one that is characterized by the minimum total cost. In other words, the hypercube $\tilde{\chi}_r$ must satisfy the solution of the following problem:

$$\min_{\tilde{\chi}_r}\left\{\sum_{\substack{1\leq i\leq n\\ x_i\in\tilde{\chi}_r}}c_ix_i\left|\frac{1}{N}\sum_{1\leq j\leq N}\delta_r^{(j)}\geq R^0\right|\right\}. \tag{8.7}$$

Now let the specified requirement be given for the system average time to failure, T^*. The hypercube $\tilde{\chi}_r'$ presenting the solution must be chosen corresponding to the solution of the problem:

$$\min_{\tilde{\chi}_r}\left\{\sum_{\substack{1\leq i\leq n \\ x_i\in\tilde{\chi}_r}} c_i x_i \left| \frac{1}{N}\sum_{1\leq j\leq N} \tau_r^{(j)} \geq T^0 \right.\right\}. \tag{8.8}$$

Of course, one should take into account that the operation with the frequency differs from the operation with the probability. The proposed solution is asymptotically accurate. Thus, all these arguments are satisfactory only for a large enough sample size.

8.5 INVERSE OPTIMIZATION PROBLEM

8.5.1 System Successful Operation versus System Cost

We need to find x^{opt} that satisfies Equation (8.5). The algorithm for solution is as follows.

Step 1. Choose a hypercube χ_1 whose diagonal vertex is lying on or under the hyperplane in Equation (8.3).

Step 2. Take the first realization, $\psi^{(1)}$, obtained with the help of the Stopping Rule 1. Find moment $\tau_1^{(1)}$ when this trajectory "punctures" the hypercube χ_1. This corresponds to the moment $t_k^{(1)}$ where

$$k = \left\{k : (\forall x_{i,k-1} \leq \chi_{i1}) \wedge (\exists x_{ik} > \chi_{i1}) \text{ IS TRUE}\right\},$$

and where χ_{i1} is a component of χ_1.

Step 3. Assign to $\tau_1^{(1)}$ a value 1 or 0 by using the indicator function in Equation (8.4).

Step 4. Add $\delta_1^{(1)}$ to the value in the counter (initial value equals 0) of successful trajectories.

Repeat the procedure from step 2 for the next realization, $\psi^{(2)}$.

After the analyzing of all trajectories, we calculate the frequency of successful trajectories, $\hat{P}_1(t_0) = \frac{1}{N}\sum_{1\leq j\leq N} \delta_1^{(j)}$

for the chosen hypercube χ_1. Then we find such a hypercube χ_K that is characterized by the maximum value of $\hat{P}_K(t_0)$ For this purpose, we can use a random search, steepest descent, or other numerical optimization procedure in the discrete space of the trajectories of spare unit expenditure.

Example 8.1

Consider a series system of $n = 3$ units. For the sake of simplicity of illustrative calculations and possibility to compare an obtained solution with analytical solution, assume that the system units are independent and $c_i = c$ for all i, $1 \le i \le 3$. Let the unit time to failure (TTF) be distributed exponentially with parameters $\lambda_1 = 1$, $\lambda_2 = 0.5$, and $\lambda_3 = 0.25$, respectively. The specified time of failure-free operation is $t_0 = 1$. Admissible total cost of a spare unit is equal to 4.

In the left three columns of Table 8.1, there are random exponentially distributed times to failure, ξ_i, for all three units. In the next three columns there are corresponding sequences of replacement times $\theta_{i(k)}$: $\theta_{i(k)} = \xi_{i1} + \xi_{i2} + \ldots + \xi_{ik}$. In other words, $\theta_{i(k)}$ is a random survival time of the ith redundant group consisting of one main and k-1 spare units.

We will use the same random numbers for all of the examples below. This leads to a dependence of the results obtained for different problems, but our main goal is to illustrate the algorithm of the solution with the use of a numerical example, rather than to execute an accurate statistical experiment.

Solution to Example 8.1

(1) Consider Realization 1 from Table 8.1. First, take values ξ_i of the first row of the left block of columns: $\xi_{11} = 0.07$, $\xi_{21} = 0.75$, and $\xi_{31} = 0.24$. Denote them $\theta_{1(1)}$, $\theta_{2(1)}$, and $\theta_{3(1)}$, respectively, and set them into the first row of the right

TABLE 8.1   Random TTF and Replacement Time for 10 Monte Carlo Realizations

TTF			Replacement time		
unit 1	unit 2	unit 3	unit 1	unit 2	unit 3
Realization 1					
0.07	0.75	0.24	0.07	0.75	0.24
0.53	4.97	3.19	0.6	5.72	3.43
0.06	0.45	1.41	0.66	6.17	4.84
0.53	2.59	3.42	1.19	8.76	8.26
1.44	5	1.59	2.63	13.76	9.85
Realization 2					
0.42	0.13	4.92	0.42	0.13	4.92
0.16	1.15	12.9	0.58	1.28	17.82
0.45	3.29	0.83	1.03	4.57	18.65
0.28	0.35	2.35	1.31	4.91	21
0.25	2.1	1.74	1.55	7.02	22.73
Realization 3					
0.62	3.47	3.22	0.62	3.47	3.22
3.66	2.72	3.92	4.28	6.2	7.14
5.11	2.47	3.9	9.39	8.67	11.04
0.31	1.69	1.21	9.7	10.36	12.25
1.42	0.86	0.96	11.12	11.22	13.21
Realization 4					
1.45	5.85	0.51	1.45	5.85	0.51
1.13	1.26	8.64	2.58	7.11	9.15
1.27	2.14	4.71	3.85	9.25	13.86
0.45	0.67	1.16	4.29	9.92	15.01
2.48	1.52	6.38	6.77	11.44	21.4
Realization 5					
0.32	0.22	0.54	0.32	0.22	0.54
0.75	0.15	1.53	1.08	0.37	2.06
0.73	1.49	1.78	1.81	1.87	3.84
0.01	0.68	0.89	1.82	2.55	4.73
0.25	3.06	1.68	2.07	5.6	6.41

TABLE 8.1   (*Continued*)

TTF			Replacement time		
unit 1	unit 2	unit 3	unit 1	unit 2	unit 3
Realization 6					
0.11	2.03	5.54	0.11	2.03	5.54
1.03	0.48	10.57	1.13	2.52	16.11
0.88	2.26	5.14	2.01	4.77	21.25
0.39	5.19	0.92	2.41	9.96	22.17
3.45	1.12	6.58	5.86	11.08	28.74
Realization 7					
1.22	0.11	2.69	1.22	0.11	2.69
1.87	0.91	0.1	3.09	1.02	2.79
0.41	2.11	1.9	3.5	3.13	4.69
3.95	0.36	3.72	7.45	3.49	8.41
0.4	1.67	1.43	7.85	5.17	9.84
Realization 8					
0.27	1.49	22.49	0.27	1.49	22.49
0.44	0.53	1.24	0.71	2.02	23.73
0.74	1.07	12.07	1.45	3.09	35.8
0.76	1.13	2.86	2.2	4.22	38.65
0.36	2.99	2.87	2.57	7.21	41.52
Realization 9					
0.46	1.55	7.9	0.46	1.55	7.9
1.06	4.8	7.59	1.52	6.35	15.49
1.9	2.66	8.14	3.42	9.01	23.63
0.17	0.37	1.26	3.59	9.38	24.89
2.18	0.43	5.17	5.77	9.8	30.06
Realization 10					
0.83	1.08	0.58	0.83	1.08	0.58
0.4	1.76	3.76	1.23	2.84	4.33
1	0.94	8.73	2.23	3.79	13.07
0.4	1.48	3.74	2.63	5.26	16.81
0.47	2.91	3.73	3.1	8.18	20.54

block of columns ("Replacement time"). Find the minimum value: min $\{\theta_{1(1)}, \theta_{2(1)}, \theta_{3(1)}\} = \theta_{1(1)} = 0.07$.

(2) Next take the value $\xi_{12} = 0.53$ in the column "TTF; Unit 1." Form a new value: $\theta_{1(2)} = \theta_{1(1)} + \xi_{12} = 0.07 + 0.53 = 0.6$. Rename $\theta_{2(1)} = \theta_{2(2)}$ and $\theta_{3(1)} = \theta_{3(2)}$. Set this value into the second place in the column "Replacement time; Unit 1."

(3) Find the minimum value: min $\{\theta_{1(2)}, \theta_{2(2)}, \theta_{3(2)}\} = \theta_{3(2)} = 0.24$. Repeat step 2 until the total cost of each system equals 7. (For the case $c_i = c$, it means that all 7 units are spent.) As the result, we spent three units of type 1, no units of type 2, and two units of type 3. In this particular case, the system TTF does not reach the specified time $t_0 = 1$.

(4) Repeat steps 1 to 3 with the remaining realizations from Table 8.1 and complete Table 8.2.

(5) Notice that units marked with "*" in Table 8.2 are auxiliary, that is, in each particular case they are not necessary because t_0 has been reached before all permitted resources were spent.

(6) In Table 8.3, list all vectors: $X^{(k)} = (x_1^{(k)}, x_2^{(k)}, x_3^{(k)}), k = 1, 2, \ldots,$ 10, that are obtained from Table 8.2 after exclusion of the marked units (see Table 8.3).

Realization #1 is not taken into account because its TTF < 1.

(7) Order each component of these vectors separately (see Table 8.4). In other words, Table 8.4 shows the frequency with which a corresponding number of spare units of each type has been met during 10 realizations of the Monte Carlo simulation. We see that the use of the vector $(3, 3, 2)$ of spare units for this realization will lead to 1 failure in 10 experiments. However, the total system cost equals 8 units. So, the next step is to find the best way of reducing the total system cost.

TABLE 8.2 Initial Experiment with Exclusion of "Extra Units" (Marked with "*")

Unit 1	Unit 2	Unit 3	Unit 1	Unit 2	Unit 3
	Realization 1			Realization 6	
0.07	0.75	0.24	0.11	2.03	5.54
0.6		3.43	1.13	2.52*	
0.66			2.01*		
1.19			2.41*		
	Realization 2			Realization 7	
0.42	0.13	4.92	1.22	0.11	2.69
0.58	1.28		3.09*	1.02	2.79*
1.03				3.13*	
1.31*				Realization 8	
	Realization 3		0.27	1.49	22.49
0.62	3.47	3.22	0.71	2.02*	
4.28	6.2*	7.14*	1.45		
9.39*			2.2*		
	Realization 4			Realization 9	
1.45	5.85	0.51	0.46	1.55	7.9
2.58*	9.15		1.52	6.35*	
3.85*			3.42*		
4.29*			3.59*		
	Realization 5			Realization 10	
0.32	0.22	0.54	0.83	1.08	0.58
1.08	0.37	2.06	1.23	2.84*	4.33
	1.87		2.23*		

(8) Put the number of realizations for which TTF has not reached $t_0 = 1$ into the failure counter. In our case there is only one such realization with TTF ≤ 1.

(9) Exclude from Table 8.2 all vectors which correspond to the realizations mentioned in step 7.

TABLE 8.3 Realization of Units Spent (Corrected for $t_0 \geq 1$)

Realization number	x_1	x_2	x_3	System's TTF
1	4*	1*	2*	<1
2	3	2	1	1.03
3	2	1	1	3.22
4	1	1	2	1.45
5	2	3	2	1.08
6	2	1	1	1.13
7	1	2	1	1.02
8	3	1	1	1.45
9	2	1	1	1.55
10	2	1	2	1.08
Maximum	**3**	**3**	**2**	

TABLE 8.4 Ordered Numbers of the Use of Units of Different Types

x_1	x_2	x_3
1	1	1
2	1	1
2	1	1
2	1	1
2	1	1
2	(1)	2
3	2	2
3	2	2
3	2	2
(4)	3	(2)

Note: Numbers in parentheses correspond to the first realization, which was not taken into account.

(10) Find which unit in Table 8.4 has the smallest number of the use of largest values of $\max_{1 \le k \le 10} x_i^{(k)}$. In our example, three units of type 1 were used in three realizations, three units of type 1 were used once, and two units of type 3 were used four times. (We exclude from consideration Realization 1 since it did not deliver TTF ≥ 1.) In this case we exclude one unit of type 2 because in this case we gain one unit of cost and "loss" only one realization.

(11) Add number of units excluded at step 9 into the counter of system failures.

(12) Check if the system cost equal to or less than $C^0 = 7$. If "No," correct Table 8.3 by exclusion of the vector 1 and continue the procedure from step 6. If "Yes," stop the procedure.

(13) Calculate the ratio of realization without failure (the total number of realization minus the number of failures from the counter) to the total number of performed realizations.

In the example considered the final solution is (3, 2, 2).

As a direct calculation using tables of Poisson distribution shows, this vector delivers the probability 0.804. Of course, such a coincidence with the observed frequency equal to 0.8 in a particular statistical experiment is not a proof of the method. However, multiple results obtained by the proposed method for other examples show a proper closeness to the exact solution, even for a relatively small sample size.

The asymptotical convergence of the solution to the optimal one was proved by Gordienko and Ushakov (1978).

8.5.2 System Average Time to Failure versus System Cost

We need to find X'_{opt} that satisfies the solution of Equation (8.6). In this case, the algorithm almost completely coincides with the one described above. The only difference is in the absence of step 3. At step 4 we directly place $\tau_1^{(i)}$ in a counter of the survival time. After analyzing all of the trajectories, the estimate of the MTTF for the hypercube χ_1 is calculated as

$$T_1 = \frac{1}{N} \sum_{1 \le i \le N} \tau_1^{(1)}. \tag{8.9}$$

After this, we perform the analogous calculations for other hypercubes, finding those which are characterized by the maximum estimate of MTTF. The search for the maximum can be performed in the same way as was done previously.

Example 8.2

We will consider the same data as in the example above. The system is again allowed to have at most 7 units in total.

Repeat steps from 1 to 4 of the process described in Section 8.3. In other words, we assume that Table 8.2 is constructed. For the solution of this problem, we will use all the data from Table 8.2. The continuation of the algorithm for this case is as follows.

Consider the vectors of Table 8.2. In this case, the components marked with "*" are included. Those vectors are obtained in the imitation process until 7 units of price are spent. Now extract corresponding values from the right side of Table 8.1 (see Table 8.5). In this table we can see how long each unit was operating.

On the basis of Table 8.5, we compose Table 8.6. In each position of this table we have the total sum of the time spent during all realizations. First of all, for independent and identical units, these values depend on the number of realizations where this unit was observed. (In the general case, where units are different and could be dependent, the number of such realizations might not be a dominant parameter.) These values from the bottom show how much we will lose by excluding a unit.

TABLE 8.5　Random Time to Failure for Each Realization until Expenditure of Seven Units

Unit 1	Unit 2	Unit 3	Unit 1	Unit 2	Unit 3
Realization 1			Realization 6		
0.07	0.75	0.24	0.11	2.03	5.54
0.53		3.19	1.03	0.48	
0.06			0.88		
0.53			0.39		
Realization 2			Realization 7		
0.42	0.13	4.92	1.22	0.11	2.69
0.16	1.15		1.87	0.91	0.1
0.45				2.11	
0.28			Realization 8		
Realization 3			0.27	1.49	22.49
0.62	3.47	3.22	0.44	0.53	
3.66	2.72	3.92	0.74		
5.11			0.76		
3.9			Realization 9		
Realization 4			0.46	1.55	7.9
1.45	5.85	0.51	1.06	4.8	
1.13		8.64	1.9		
1.27			0.17		
0.45			Realization 10		
Realization 5			0.83	1.08	0.58
0.32	0.22	0.54	0.4	1.76	3.76
0.75	0.15	1.53	1		
	1.49				

TABLE 8.6 Sum of the Times Spent by Units on the Specified Positions

Unit 1	Unit 2	Unit 3
5.77	16.68	48.63
11.03	12.5	21.14
11.41	3.6*	3.9*
2.58*		

*Units that are eliminated ($x_1 = 4$, $x_2 = 3$, and $x_3 = 3$).

It is clear that the loss will be less if we leave $x_1 = 3$, $x_2 = 2$, and $x_3 = 2$. By eliminating them, we can decrease the total system cost by up to 7 units of price.

The time to failure for the system in each realization is calculated as the minimal value among those, which are restricted by vector ($x_1 = 3$, $x_2 = 2$, $x_3 = 2$), that is, for the kth realization,

$$\xi^{(k)} = \min\{\xi_{13}^{(k)}, \xi_{22}^{(k)}, \xi_{32}^{(k)}\} \quad \xi_{Syst}^{(k)} = \min(\xi_{13}^{(k)}, \xi_{22}^{(k)}, \xi_{32}^{(k)}).$$

These values can be found in the right-hand column of Table 8.6. The results are shown in Table 8.7. These values allow calculating the mean time to failure of the investigated system.

8.6 DIRECT OPTIMIZATION PROBLEM

8.6.1 System Cost versus Successful Operation

We need to find $\tilde{\chi}_r$ which satisfies the solution of Equation (8.5). The algorithm of solution in this case is as follows.

> **Step 1.** Construct a realization of the first trajectory of the spare unit expenditure until $t_1^{(1)}$ exceeds the specified

TABLE 8.7 Time to Failure for 10 Realizations Picked up for Vector (3, 2, 2)

Realization number	TTF
1	0.66
2	1.03
3	6.2
4	3.85
5	0.37
6	2.01
7	1.02
8	1.45
9	3.42
10	2.23

value of operational time t_0. Memorize the number of spare units spent, $x_i^{(1)}$, $i = 1, 2, \ldots, n$. Continue this procedure until all of N required trajectories are constructed.

Step 2. Construct a hypercube χ_1 whose edges χ_{i1} are found as $\chi_{i1} = \max_{1 \le j \le N} x_{ij}$ that is, χ_{i1} is the maximum number of spare units of type i observed during all N realizations. (It means that for this particular sample of realizations, all of them will lay within a hypercube that is stocked with enough spare units so that we would not observe any system failure.)

Step 3. Calculate the system cost for the hypercube $\chi^{(1)}$ for which all trajectories have the survival time no less than t_0:

$$C_{\max} = \sum_{1 \le i \le n} c_i \chi_{i1}.$$

Step 4. Calculate for each i:

$$\gamma_i^{(1)} = \frac{v_i^{(1)}}{\Delta c_i^{(1)}}.$$

where $v_i^{(1)}$ shows how many numbers equal to χ_{i1} exist for a unit of type i and $\Delta c_i^{(1)}$ is the value of the system cost decrease if we reject to use $\max_{1 \leq j \leq N} \chi_{ij}$ and will use the next value in the descending order.

Step 5. Find the type of units that correspond to the maximum value of $\gamma_i^{(1)}$ and name it as i_1, that is, this number corresponds to the following condition:

$$i_1 = \left\{ i : \gamma_i = \max_{1 \leq j \leq n} \gamma_j^{(1)} \right\}.$$

Step 6. Exclude $v_{i_1}^{(1)}$ units of type i_1 and form a new value:

$$\chi_{i_1 2} = \chi_{i_1 1} - v_{i_1}^{(1)}.$$

Step 7. Rename remaining numbers:

$$\chi_{i_1 2} = \chi_{i_1 1} \text{ for all } i_j \neq i_1.$$

Step 8. Calculate the system successful operation index after the exclusion of v_{i_1} units of type i_1:

$$\hat{P}^{(2)} = 1 - \frac{v_{i_1}^{(1)}}{N}.$$

Step 9. Calculate the system spare units' cost after the exclusion of v_{i_1} units of type i_1:

$$C^{(2)} = C_{\max} - c_{i_1} v_{i_1}^{(1)}.$$

After these steps we have a new hypercube χ_2:

$$\chi_2 = \{ \chi_{12}, \chi_{22}, \dots, \chi_{n2} \}.$$

Repeat the procedure from step 5 until the system spare units' cost is equal to or smaller than the given restriction.

Example 8.3

Let us take $\hat{R}(X) \geq 0.9$. In the previous example we found that the vector of spare units (4, 3, 2) satisfies 100% of successful realizations. So, if we take a vector (3, 3, 2) it will satisfy the condition $\hat{R}^0 = 0.9$ Now we need to find the lower 90% confidence limit for the frequency 0.9 obtained in 10 experiments. This limit can be found with the use of the Clopper-Pearson method.

In this particular case it is easier to make direct calculations. If we choose the estimate of the searched probability equal to 0.9 then the probability that we will observe no fewer than 8 successes is equal to:

$$0.9^{10} + \binom{10}{1}(0.1)(0.9)^9 + \binom{10}{2}(0.1)^2(0.9)^8 = 0.9298.$$

So, in the process of decreasing the number of used spare units, we must stop after the first exclusion, that is, the solution in this case is (3, 3, 2).

Thus, the solutions of direct and inverse problems of optimal redundancy are different, though they should coincide. The variation lies in the difference of approaches: having the restriction on the system cost, we maximize the possible observed *frequency*; in the latter case we consider minimization of the system cost under the condition that the level of *probability* is guaranteed. This difference will be smaller if the number of realizations is larger.

We could solve the problem above with an iteration procedure using the solution of the direct problem of optimal redundancy. The use of the "fork method" is convenient in this case. We find the solution, X_1^{opt} for some cost restriction, say, C_1^*, and calculate the value $R_1 = R(\chi_1^{opt})$. If $R_1 < R^*$ we chose $C_2^* > C_1^*$ and continue the procedure; if $R_1 > R^*$ we chose $C_2^* < C_1^*$ and also continue the procedure. For the next steps, we can use a simple linear approximation:

$$C_{k+1}^* = C_k^* - (C_k^* - C_{k-1}^*)\frac{R_k - R^*}{R_k - R_{k-1}}.$$

where subscript k stands for the current step, subscript $k-1$ for the previous step, and subscript $k+1$ for the next step.

8.6.2 System Cost versus Average Time to Failure

We need to find $\tilde{\chi}_{r'}$, which satisfies the solution of Equation (8.6). We could not find a convenient procedure for solving this particular problem. One might consider using an interactive procedure using the sequential solution of the second direct problem considered above. For instance, we can fix some restriction on the system cost, say, $C_{syst}^{(1)}$, and find the corresponding optimal solution for $\hat{T}_{syst}^{(1)}$. If this value is smaller than the required $\hat{T}_{syst}^*$, it means that the system cost must be increased, say, up to some $C_{syst}^{(2)} > C_{syst}^{(1)}$. If $\hat{T}_{syst}^{(1)} > \hat{T}_{syst}^*$, one must choose $C_{syst}^{(2)} < C_{syst}^{(1)}$. This procedure continues until a satisfactory solution is obtained. At an intermediate step L for choosing $\hat{T}_{syst}^{(L)}$, one can use the linear extrapolation method. For example, assume that in first situation described above, the value $\hat{T}_{syst}^{(2)}$ is still less than $\hat{T}_{syst}^*$. Then the value of $C_{syst}^{(3)}$ can be chosen from the following equation:

$$\frac{C_{syst}^{(3)} - C_{syst}^{(1)}}{C_{syst}^{(2)} - C_{syst}^{(1)}} = \frac{T_{syst}^* - T_{syst}^{(1)}}{T_{syst}^{(2)} - T_{syst}^{(1)}}.$$

Obviously, one can also use the procedure similar to that in a solution of the direct problem. However, one should somehow find an initial hypercube and construct all trajectories within it. (There is no stopping rule in this case.) Then one should construct a system of embedded hypercubes and again use the steepest descent.

While solving this problem one must remember that the condition $\hat{T}_{syst} \geq T^*$ can be considered only in a probabilistic sense.

CHRONOLOGICAL BIBLIOGRAPHY

Ushakov, I. A., and Yasenovets, A. V. 1977. "Statistical methods of solving problems of optimal standby." *Soviet Journal of Computer and System Sciences*, no. 6.

Ushakov, I. A., and Gordienko, E. I. 1978. "On statistical simulation approach to solution of some optimization problems." *Elektronische Informationsverarbeitung und Kybernetik*, no. 3.

Ushakov, I. A., and Gordienko, E. I. 1978. "Solution of some optimization problems by means of statistical simulation." *Electronosche Infdormationsverarbeitung und Kybernetik*, no. 11.

Mohan, C., and Shanker, K. 1988. "Reliability optimization of complex systems using random search technique." *Microelectronics and Reliability*, no. 28.

Boland, P. J., El-Neweihi, E., and Proschan, F. 1992. "Stochastic order for redundancy allocations in series and parallel systems." *Advances in Applied Probability*, no. 1.

Zhao, R., and Liu, B. 2003. "Stochastic programming models for general redundancy-optimization problems." *IEEE Transaction on Reliability*, no. 52.

CHAPTER 9

COMMENTS ON CALCULATION METHODS

9.1 COMPARISON OF METHODS

Optimal redundancy is a very important practical problem. The solution of the problem allows one to improve reliability at a minimal expense. But here, as in many other practical problems, questions arise: What is the accuracy of the obtained results? What is the real effect of the use of sophisticated mathematics?

These are not unreasonable questions.

We have already discussed what it means to design an "accurate" mathematical model. It is always better to speak about a mathematical model which more or less correctly reflects a real object. But let us suppose that we are "almost sure" that the model is perfect. What price are we willing to pay for obtaining numerical results? What method is best, and best in what sense?

The use of excessively accurate methods is, for practical purposes, usually not necessary because of the uncertainty of the sta-

Optimal Resource Allocation: With Practical Statistical Applications and Theory, First Edition. Igor A. Ushakov. © 2013 John Wiley & Sons, Inc. Published 2013 by John Wiley & Sons, Inc.

tistical data. On the other hand, it is inexcusable to use approximate methods without reason.

We compare the different methods in the sense of their accuracy and computation complication.

The Lagrange multiplier method (LMM) demands the availability of continuous, differentiable functions. This requirement is met very rarely: one usually deals with the essentially discrete nature of the resources. But LMM sometimes can be used for a rough estimation of the desired solution.

The steepest descent method (SDM) is very convenient from a computational viewpoint. It is reasonable to use this method if the resources that one might spend on redundancy are large. Of course, this generally coincides with the requirement of high system reliability because this usually involves large expenditures of resources. But unfortunately, it happens very rarely in practice. At any rate, one can use this approach for solution of most practical problems without hesitation.

The absolute difference between costs of the two neighboring SDM solutions cannot exceed the cost of the most expensive unit value. Thus, it is clear that the larger the total cost of the system, the smaller the relative error of the solution.

The dynamic programming method (DPM) and its modifications (Kettelle's Algorithm and the method of universal generating function) are exact, but they demand more calculation time and a larger computer memory. As with most discrete problems requiring an enumerating algorithm, these optimal redundancy problems are *np*-hard.

As we mentioned above, the SDM may even provide an absolutely exact solution, since a dominating sequence for SDM is a subset of dominating sequence of DPM. In Figure 9.1, one finds two solutions obtained by SDM and DPM. The black dots correspond to the sequence obtained by SDM, and the white dots correspond to DPM.

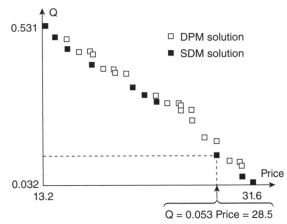

FIGURE 9.1 Comparison of DPM and SDM solutions.

Of course, one of the questions of interest is the stability of the solutions. How does the solution depend on the accuracy of the input data? How are the solutions obtained by the different methods distinguished? How much do the numerical results of the solutions differ from one method to another?

An illustration of the problem is given by numerical experiments.

Example 9.1

Consider a series system consisting of three units. The input data are assumed to be uncertain: units' probability of failure-free operation (PFFO) and cost are known with an accuracy of 10%. To demonstrate possible difference in solutions, let us take the five systems presented in Table 9.1.

The problem is to check the stability of the optimal solutions over the range of variability of the parameters.

TABLE 9.1 Five Different Systems Consisting of Similar Units with Slightly Different Values of Parameters

System	Unit 1		Unit 2		Unit 3	
	q	c	q	c	q	c
A	0.2	1	0.2	1	0.2	1
B	0.2	0.9	0.2	1	0.2	1.1
C	0.18	0.9	0.2	1	0.22	1.1
D	0.18	1.1	0.2	0.1	0.22	0.9
E	0.18	1	0.2	1	0.22	1

TABLE 9.2 Optimal Numbers of Redundant Units for Five Systems Presented in Table 9.1 (the Level of Spent Resources Is under 30 Units)

System	Number of redundant units			Probability of system failure	Factual system cost
	x_1	x_2	x_3		
A	10	10	10	$3.07 \cdot 10^{-7}$	30
A*	10	10	10	$3.07 \cdot 10^{-7}$	30
B	10	10	10	$3.07 \cdot 10^{-7}$	30
B*	10	10	10	$3.07 \cdot 10^{-7}$	30
C	9	10	10	$5.66 \cdot 10^{-7}$	29.1
C*	10	10	10	$4.04 \cdot 10^{-7}$	30
D	9	10	11	$3.59 \cdot 10^{-7}$	29.8
D*	9	10	11	$3.59 \cdot 10^{-7}$	29.8
E	9	10	11	$3.59 \cdot 10^{-7}$	29.8
E*	9	10	11	$3.59 \cdot 10^{-7}$	29.8

Solution to Example 9.1

At first, we compare the solutions for all five cases where the specified total system cost is to be at most 30 units. For each case, we give two results: one obtained by the SDM and the second (marked with *) obtained by the DPM. The results are presented in Table 9.2.

TABLE 9.3 Optimal Numbers of Redundant Units for Five Systems Presented in Table 9.1 (the Level of Required Reliability Is under 1·10 in Power)

System	Number of redundant units			Probability of system failure	Factual system cost
	x_1	x_2	x_3		
A	9	10	10	$7.7 \cdot 10^{-7}$	30
	Equivalent solutions are (10,9,10) and (10,10,9)				
A*	9	10	10	$7.17 \cdot 10^{-7}$	30
B	10	10	9	$7.17 \cdot 10^{-7}$	30
B*	10	10	9	$7.17 \cdot 10^{-7}$	30
C	9	9	10	$9.76 \cdot 10^{-7}$	29.1
C*	9	9	10	$9.76 \cdot 10^{-7}$	30
D	9	9	10	$9.76 \cdot 10^{-7}$	29.8
D*	9	9	10	$9.76 \cdot 10^{-7}$	29.8
E	9	9	10	$9.76 \cdot 10^{-7}$	29.8
E*	9	9	10	$9.76 \cdot 10^{-7}$	29.8

The table shows that the only differences between the approximate and exact solutions are observed for the cases C and C*. However, all solutions are very close.

With an increase in spent resources, the relative difference between the solutions obtained by the SDM and the DPM will be increasingly smaller.

We now analyze the solutions corresponding to a specified level of reliability. In Table 9.3, for the same systems, respective results for $Q_0 = 1 \cdot 10^{-6}$ are collected.

Numerical computer experiments and practical experience in solution of the optimal redundancy problem could aid in the development of a keen engineering intuition in the approximate solving of such problems and their sensitivity analysis.

9.2 SENSITIVITY ANALYSIS OF OPTIMAL REDUNDANCY SOLUTIONS

Solving practical optimal redundancy problems, one can ponder: what is the sense of optimizing if input data are plucked from the air? Indeed, statistical data are so unreliable (especially in reliability problems) that such doubts have a very good ground.

Not finding any sources after searching for the answer to this question, the author decided to make some investigation of optimal solution sensitivity under influence of data scattering.

A simple series system of six units has been considered (see Fig. 9.2). For reliability increase, one uses a loaded redundancy, that is, if a redundant group k has x_k redundant units, its reliability is

$$P_k(x_k) = 1 - (1 - p_k)^{x_k+1},$$

where p_k is the PFFO of a single unit k. The total cost of x_k redundant units is equal to $c_k \cdot x_k$, where c_k is the cost of a single unit of type k.

Units' parameters are presented in Table 9.4.

Assume that units are mutually independent, that is, the system's reliability is defined as:

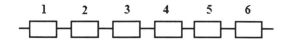

FIGURE 9.2 Series system that underwent analysis.

TABLE 9.4 Input Data

	Unit 1	Unit 2	Unit 3	Unit 4	Unit 5	Unit 6
1. p_k	0.8	0.8	0.8	0.9	0.9	0.9
2. c_k	5	5	5	1	1	1

$$P_{System}(x_k, 1 \le k \le 6) = \prod_{1 \le k \le 6} P_k(x_k)$$

and the total system cost is:

$$C_{System}(x_k, 1 \le k \le 6) = \sum_{1 \le k \le 6} c_k x_k$$

Solutions of both problems of optimal redundancy are presented here:

direct: $\min_{1 \le x_k < \infty} \{C(x_k, 1 \le k \le 6) | P(x_k, 1 \le k \le 6) \ge P^*\}$

and

inverse: $\max_{1 \le x_k < \infty} \{P(x_k, 1 \le k \le 6) | C(x_k, 1 \le k \le 6) \le C^*\}.$

For finding the optimal solutions, the steepest descent method was applied. For this "base" system, the solutions for several sets of parameters are presented for the direct problem in Table 9.5 and for the inverse problem in Table 9.6. (Numbers are given with high accuracy only for demonstration purposes; in practice, one has to use only significant positions after a row of nines.)

The question of interest is: How will the optimal solution change if input data are changed? Two types of experiments have been performed: in the first series of experiments, different units'

TABLE 9.5   Solution for the Direct Problem

P^*	x_1	x_2	x_3	x_4	x_5	x_6	Achieved P	System C
0.95	3	3	3	3	2	2	0.9559520	52
0.99	4	4	3	3	3	3	0.991187	69
0.995	5	4	4	4	3	3	0.995229	75
0.999	6	5	5	4	4	4	0.999218	93

TABLE 9.6   Solution for the Inverse Problem

C^*	x_1	x_2	x_3	x_4	x_5	x_6	Achieved C	System P
50	3	3	2	2	2	2	46	0.931676
75	4	4	3	3	3	3	75	0.995229
100	5	4	4	4	3	3	99.5	0.999602

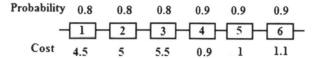

FIGURE 9.3   Input data for the first series of experiments.

FIGURE 9.4   Input data for the second series of experiments.

TABLE 9.7   Values of Probabilities of Failure-Free Operations

	0.999	0.995	0.99	0.95
Initial	0.999218	0.99566	0.9922	0.955952
Various C	0.998996	0.99566	0.9922	0.955952
Various P	0.999218	0.99566	0.9922	0.955952

costs with fixed probabilities were considered (see Fig. 9.3), and, in another one, different units' probabilities with fixed costs were considered (see Fig. 9.4).

The results of calculations are as shown in Table 9.7. In addition, a Monte Carlo simulation was performed where parameters of the PFFO and cost were changed simultaneously. In this case, parameters of each unit were calculated (in Excel) as:

$$p_k = 0.8p_k + 0.4p_k * \text{RAND}()$$

and

$$c_k = 0.8c_k + 0.4 * \text{RAND}(),$$

that is, considered a random variation of the values within ±20% limits.

The final results for this case are presented in Table 9.8, Table 9.9, Table 9.10, and Table 9.11. Analysis of data presented in these tables shows relatively significant difference in numerical results (see Fig. 9.5). However, the problem is not in coincidence of final values of PFFO or cost. The problem is how the change of parameters influences the optimal values of $x_1, x_2, \ldots$.

One can observe that even with a system of six units (redundant groups), a visual analysis of sets $(x_1, x_2, \ldots, x_6)$ is extremely difficult, and, at the same time, deductions based on some averages or deviations of various x_k are almost useless.

TABLE 9.8   Results of Monte Carlo Simulations for $P^* = 0.999$

No.	$P^* = 0.999$							
	P	C	x_1	x_2	x_3	x_4	x_5	x_6
1	0.999352	100	6	6	6	4	4	4
2	0.999218	102	6	6	6	5	4	4
3	0.999313	102	6	6	6	4	4	4
4	0.999212	97	5	6	6	4	4	4
5	0.999182	102	6	6	6	4	4	4
6	0.999171	97	6	6	5	4	4	4
7	0.999171	103	6	6	6	4	5	4
8	0.999596	100	6	6	6	4	4	4
9	0.999526	100	6	6	6	4	4	4
10	0.999399	100	6	6	6	4	4	4

TABLE 9.9   Results of Monte Carlo Simulations for $P^* = 0.995$

No.	$P^* = 0.995$							
	P	C	x_1	x_2	x_3	x_4	x_5	x_6
1	0.995478	84	5	5	5	3	3	3
2	0.996755	85	5	4	4	4	3	3
3	0.995026	85	5	4	5	4	3	3
4	0.996777	79	4	5	5	3	3	3
5	0.996777	84	5	5	5	3	3	3
6	0.995525	79	5	5	4	3	3	3
7	0.996732	85	5	5	5	3	4	3
8	0.996732	85	5	5	5	3	4	3
9	0.995645	84	5	5	5	3	3	3
10	0.99567	84	5	5	5	3	3	3

TABLE 9.10   Results of Monte Carlo Simulations for $P^* = 0.99$

No.	$P^* = 0.99$							
	P	C	x_1	x_2	x_3	x_4	x_5	x_6
1	0.990147	69	4	4	4	3	3	3
2	0.990965	70	4	4	4	4	3	3
3	0.990229	70	4	4	4	4	3	3
4	0.99185	69	4	4	4	3	3	3
5	0.990389	71	4	4	4	4	4	3
6	0.99107	69	4	4	4	3	3	3
7	0.992185	74	5	4	4	3	3	3
8	0.990422	71	4	4	4	3	4	3
9	0.990893	71	5	4	4	3	3	3
10	0.990466	69	4	4	4	3	3	3

TABLE 9.11   Results of Monte Carlo Simulations for $P^* = 0.95$

No.	$P^* = 0.95$							
	P	C	x_1	x_2	x_3	x_4	x_5	x_6
1	0.950045	52	3	3	3	3	2	2
2	0.955842	52	3	3	3	3	2	2
3	0.951936	52	3	3	3	3	2	2
4	0.951711	54	3	3	3	2	2	2
5	0.957883	50	3	3	3	3	3	2
6	0.951908	51	3	3	3	2	2	2
7	0.962227	51	3	3	3	2	2	2
8	0.962227	51	3	3	3	3	3	2
9	0.95261	50	3	3	3	3	2	3
10	0.950393	52	3	3	3	3	2	2

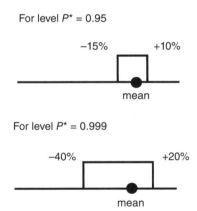

For level $P^* = 0.95$

−15%   +10%

mean

For level $P^* = 0.999$

−40%   +20%

mean

FIGURE 9.5   Deviation of maximum and minimum values of probability of failure-free operation obtained by Monte Carlo simulation.

The author was forced to invent some kind of a special presentation of sets of x_ks. Since there is no official name for this kind of graphical presentation, it will be called a "multiple polygon." On this type of multiple polygon, there are numbers of "rays" corresponding to the number of redundant of units (groups). Each ray has several levels corresponding to the number of calculated redundant units for the considered case (see Fig. 9.6).

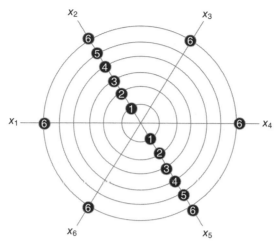

FIGURE 9.6 Multiple polygon axes with numbered levels.

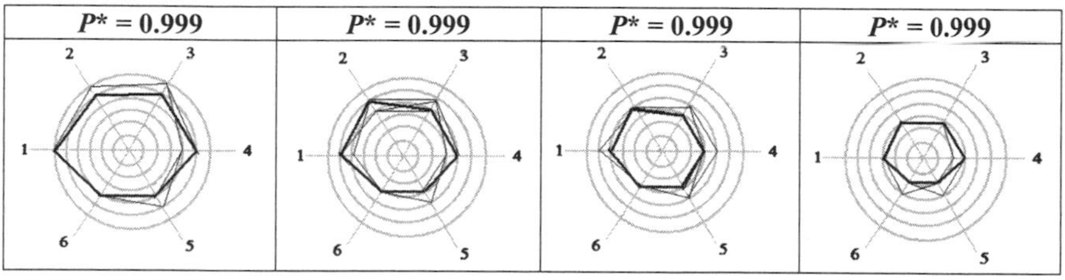

FIGURE 9.7 Deviations of optimal solutions for randomly varied parameters from the optimal solution obtained for parameters given in Table 9.1.

The multiple polygons give a perfect visualization of "close-to-optimal" solutions and characterize observed deviation of particular solutions. Such multiple polygons for considered example are given in Figure 9.7. (Here, bold lines are used for connecting the values of x_k obtained as optimal solution for units with parameters given in Table 9.4.)

Thus, one may notice that input parameter variation may influence significantly enough the probability of failure-free operation and the total system cost from run to run of Monte Carlo simulation, though the optimal solution remains more or less stable.

CHAPTER *10*

OPTIMAL REDUNDANCY WITH SEVERAL LIMITING FACTORS

10.1 METHOD OF "WEIGHING COSTS"

A number of cases arise where one has to take into account several restrictions in solving the optimal redundancy problem. For example, various objects such as aircraft, satellites, and submarines have restrictions on cost and also on weight, volume, required electric power, and so on. (Apparently, the cost for most of these technical objects is an important factor, but, perhaps, less important than others mentioned.)

In these cases, one has to solve the optimization problem under several restrictions and to maximize the system reliability index under restrictions on all other factors.

Consider a system consisting of n redundant groups connected in series. For each additional redundant unit of the system, one has to spend some quantity of M various types of resources (for instance,

Optimal Resource Allocation: With Practical Statistical Applications and Theory,
First Edition. Igor A. Ushakov.
© 2013 John Wiley & Sons, Inc. Published 2013 by John Wiley & Sons, Inc.

cost, weight, or volume), say, $C_j(X)$. There are constraints on each type of resource: $C_j(X) \le C_j^0$. The optimization problem is formulated as:

$$\max_{X}\{R(X)|C_1(X) \le C_1^0, C_2(X) \le C_2^0, \ldots, C_M(X) \le C_M^0\}, \quad (10.1)$$

where $X = (x_1, x_2, \ldots, x_n)$ is the vector of the system redundant units.

Let us assume that each $C_j(X)$ is a linear function of the form

$$C_j(X) = \sum_{1 \le i \le M} c_{ji} x_i, \quad (10.2)$$

where c_{ji} is the resource of type j associated with a unit of type i.

One of the most convenient ways to solve this problem is reducing it to a one-dimensional problem. To this end, we introduce "weight" coefficients d_j such that: $0 \le d_j \le 1$, and

$$\sum_{1 \le j \le M} d_j = 1. \quad (10.3)$$

A set of d_j satisfying Equation (10.3) presents a diagonal hyperplane within n-dimensional unitary hypercube. Denote this hyperplane by D. An explanation is given for a three-dimensional case in Figure 10.1.

Use the steepest descent method (SDM) for the solution. The process of a solution is as follows. Choose a point $D^k = (d_1^k, d_2^k, \ldots, d_M^k)$, $D^k \in D$. Produce for each unit j "weighed cost" c_j^k corresponding to vector D^k:

$$c_j^k = \sum_{1 \le k \le M} c_j d_j^k. \quad (10.4)$$

Solve a one-dimensional problem while simultaneously controlling all M constraints. As soon as the optimization procedure has been stopped due to a possible violation of at least one of the

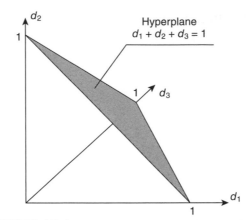

FIGURE 10.1   Hyperplane with all possible d_j.

constraints, the value of a reached level of R^k and realized vector X^k are memorized.

Then the next vector, say, D^j, is chosen and new values R^j and X^j are found. Compare admissible solutions and keep that with the largest value of R.

The procedure of the oriented choosing of D^k rather than direct enumerating can be organized: the procedure of the steepest descent could be used for this purpose.

Perhaps an even better procedure is as follows. At the stopping moment, pay attention to the type of constraint that is closest to violation. Sometimes it means that the process is "less optimal" relating to this type of resource. Increase a corresponding weight multiplier and repeat the process.

As in most practical cases, finding the appropriate choice of increment to change d is more of an art than a science.

Maximum found value R corresponds to the optimal solution X^{opt}.

An illustrative example might be useful to demonstrate this method.

EXAMPLE 10.1

Consider a series system consisting of three units with the characteristics given in Table 10.1.

A "hot" redundancy is permitted to improve the system reliability. The problem is to find the optimal solutions for the following constraints on the redundant system units as a whole:

(1) Cost: $C_1^0 = 15$ conditional units; weight: $C_2^0 = 15$ conditional units.

(2) Cost: $C_1^0 = 20$ conditional units; weight: $C_2^0 = 15$ conditional units.

Solution to Example 10.1

Choose the increment for each d_i equal to 0.25. Then the "weighed cost" can be calculated as:

$$c_\Sigma^1 = c_i^{(1)}, c_\Sigma^{0.75} = 0.75c_i^{(1)} + 0.25c_i^{(2)}, c_\Sigma^{0.5} = 0.5c_i^{(1)} + 0.5c_i^{(2)},$$
$$c_\Sigma^{0.25} = 0.25c_i^{(1)} + 0.75c_i^{(2)}, c_\Sigma^0 = c_i^{(2)}. \tag{10.5}$$

Using Equation (10.5), we get the following values for the equivalent costs:

$$c_1^2 = 2.5, c_2^2 = 4.0, c_3^2 = 2.25;$$
$$c_1^3 = 2.0, c_2^3 = 3.0, c_3^3 = 2.5;$$
$$c_1^4 = 1.5, c_2^4 = 2.0, c_3^4 = 2.75;$$
$$c_1^5 = 1.0, c_2^5 = 1.0, c_3^5 = 3.0.$$

TABLE 10.1   Data for Example 10.1

Unit (i)	Reliability index P_i	Cost C_{i1}	Weight C_{i1}
1	0.7	3	1
2	0.8	5	1
3	0.9	2	3

Now we separately solve all five problems for different equivalent costs. For simplicity, let us use the SDM. We omit all intermediate calculations that are routine and present only step-by-step results of the solution process.

Admissible solutions are (2, 2, 1) and (4, 2, 0). Solution (3, 2, 0) is not taken into account since it is dominated by (4, 2, 0). Let us now compare solutions:

$$R(2, 2, 1) = (1 - 0.3^3)(1 - 0.2^3)(1 - 0.1^2) = 0.9956$$
$$R(4, 2, 0) = (1 - 0.3^5)(1 - 0.2^3) \cdot 0.9 = 0.8906.$$

Thus the solution of the problem is vector (2, 2, 1), that is, $x_1 = 2$, $x_2 = 2$, and $x_3 = 1$.

The inverse problem of optimal redundancy occurs extremely rarely in practice, so we omit it from consideration.

10.2 METHOD OF GENERALIZED GENERATING FUNCTIONS

The problem treated above can be solved exactly with the use of the method of generalized generating functions. The legion for each ith redundant group is represented as the set of the cohorts

$$L_i = \{C_{i1}, C_{i2}, \dots, C_{iN_i}\},$$

where N_i is the number of cohorts in this legion. (In principle, the number of cohorts is unrestricted in this investigation.) Each cohort consists of $M + 2$ maniples:

$$C_{ik} = (R_{ik}, c_{ik}^1, \dots, c_{ik}^M, x_{ik})$$

where M is the number of restrictions. All maniples are defined as in the one-dimensional case that we considered above. A similar interaction is performed with the maniples:

TABLE 10.2 Solution Process for Various d_j

d_k		1	2	3	4	5	6	7	8	9	10
1	X	1,0,0	1,0,1	1,1,1	2,1,1	2,2,1	3,2,1	3,2,2	4,2,2	4,3,2	5,3,2
	C_1	3	5	6	9	10	13	16	19	20	23
	C_2	1	4	9	10	15	16	17	18	23	24
0.75	X	0,1,0	1,1,0	1,2,0	2,2,0	3,2,0	3,3,0	4,3,0	4,3,1	5,3,1	6,3,1
	C_1	1	4	5	8	11	12	15	17	20	23
	C_2	5	6	11	12	13	18	19	22	23	24
0.5	X	0,1,0	1,1,0	1,2,0	2,2,0	3,2,0	4,2,0	5,2,0	5,2,1	6,2,1	6,3,1
	C_1	1	4	5	8	11	14	17	19	22	23
	C_2	5	6	11	12	13	14	15	18	19	24
0.25	X	1,0,0	1,1,0	2,1,0	2,2,0	3,2,0	4,2,0	5,2,0	5,3,0	6,3,0	7,3,0
	C_1	3	4	7	8	11	14	17	18	21	24
	C_2	1	6	7	12	13	14	15	20	21	22
0	X	1,0,0	1,1,0	2,1,0	3,1,0	3,2,0	4,2,0	5,2,0	6,2,0	6,3,0	7,3,0
	C_1	3	4	7	10	11	14	17	20	21	24
	C_2	1	6	7	8	13	14	15	16	21	20

$$\Omega_{c^k}^M(c_{ij_i}^k, c_{lj_l}^k) = c_{ij_i}^k + c_{lj_l}^k \quad \text{and} \quad \Omega_{c^k}^M c_i^k = \sum_{1 \le i \le n} c_{ij_i}^k$$

$$\Omega_R^M(R_{ij}, R_{kl}) = R_{ij} R_{kl} \quad \text{and} \quad \Omega_R^M_{1 \le i \le n} R_{ij_i} = \prod_{1 \le i \le n} R_{ij_i} ;$$

$$\Omega_x^M(x_{ij_i}, x_{lj_l}) = (x_{ij_i}, x_{lj_l}) \quad \text{and} \quad \Omega_x^M_{1 \le i \le n} x_{ij_i} = (x_{1j_1}, x_{2j_2}, \dots, x_{nj_n}) = X_J$$

where J is the set of subscripts: $J = (j_1, \dots, j_n)$.

The remaining formal procedures totally coincide with the one-dimensional case with one very important exception: instead of a scaler ordering, one must use the special ordering of the cohorts of the final legion.

It is difficult to demonstrate the procedure on a numerical example, so we give only a detailed verbal explanation.

Suppose we have the file of current cohorts ordered according to increasing R. If we have a specified set of restrictions: $C_j(X) \le C_{0j}$ for all j: $1 \le j \le M$, then there is no cohort in this file that violates at least one of these restrictions. When a new cohort, say, C_k, appears during the interaction procedure, it is put in the appropriate place in accordance with the value of its R-maniple. The computational problem is as follows.

1. Consider a part of the current file of cohorts for which the values of their R-maniple are less than the analogous value for C_k. If, among the existent cohorts, there is a cohort, say, C^*, for which all costs are larger than those of C_k, this cohort C^* is excluded from the file.

2. Consider a part of the current file of cohorts for which the values of the R-maniple are larger than the analogous value for C_k. If between the existent cohorts there is a cohort, say, C^{**}, for which all costs are smaller than those of C_k, the new cohort is not included in the file.

3. If neither 1 nor 2 take place, the new cohort is simply added to the file in the appropriate place.

After a multi-dimensional undominated sequence is constructed, one easily finds the solution for the multiple restrictions: it is the cohort with the largest R-maniple value (in other words, a cohort on the right if the set is ordered by the values of R).

The stopping rule for this procedure is to find the size of each cohort that will produce a large enough number of cohorts in the resulting legion so as to contain the optimal solution. At the same time, if the numbers of cohorts in the initial legions are too large, the computational procedure will take too much time and will demand too large a memory space.

Of course the simplicity of this description should not be deceptive. The problem is very bulky in the sense that the multi-dimensional restrictions and the large numbers of units in typical practical problems could require a huge memory and computational time. (But who can find a non-trivial multi-dimensional problem that has a simple solution?)

CHRONOLOGICAL BIBLIOGRAPHY

Proschan, F., and Bray, T. A. 1970. "Optimum redundancy under multiple constraints." *Operations Research*, no. 13.

Ushakov, I. A. 1971. "Approximate solution of optimal redundancy problem for multipurpose system." *Soviet Journal of Computer and System Science*, no. 2.

Ushakov, I. A. 1972. "A heuristic method of optimization of the redundancy of multipurpose system." *Soviet Journal of Computer and System Sciences*, no. 4.

Nakagawa, Y., and. Miyazaki, S. 1981. "Surrogate constraints algorithm for reliability optimization problem with two constraints." *IEEE Transaction on Reliability*, no. 30.

Genis, Y. G., and Ushakov, I. A. 1984. "Optimization of multi-purpose systems." *Soviet Journal of Computer and System Sciences*, no. 3.

CHAPTER 11

OPTIMAL REDUNDANCY IN MULTISTATE SYSTEMS

Solving the problems of optimal redundancy allocation for multi-state systems (MSS) consisting of multistate units is more laborious than solving analogous problems for systems that have only two states: normal operation and failure.

Today, this problem is investigated in detail. To begin with, the works of Gregory Levitin and Anatoly Lisnjanskij have to be mentioned (a complete list of their papers is presented in the Bibliography to the chapter).

To gain a clearer perspective of this type of problem, we begin with a simple numerical example. Consider a series system of two different multistate units, each of which is characterized by several levels of performance. Performance may be measured by various physical values. Effectiveness of such system operation depends on the levels of performance of Unit 1 and Unit 2. These units are characterized by the parameters presented in Table 11.1 and Table 11.2.

Optimal Resource Allocation: With Practical Statistical Applications and Theory,
First Edition. Igor A. Ushakov.
© 2013 John Wiley & Sons, Inc. Published 2013 by John Wiley & Sons, Inc.

TABLE 11.1   Characterization of Unit 1

Level of performance (W_1)	Probability p_1	Cost of a single unit
100%	$p_{11} = \Pr\{W_1 = 100\%\} = 0.9$	$c_1 = 1$
70%	$p_{12} = \Pr\{W_1 = 100\%\} = 0.05$	
40%	$p_{13} = \Pr\{W_1 = 100\%\} = 0.04$	
0%	$p_{14} = \Pr\{W_1 = 100\%\} = 0.01$	

TABLE 11.2   Characterization of Unit 2

Level of performance (W_1)	Probability p_2	Cost of a single unit
100%	$p_{21} = \Pr\{W_2 = 100\%\} = 0.8$	$c_2 = 2$
80%	$p_{22} = \Pr\{W_2 = 80\%\} = 0.18$	
20%	$p_{23} = \Pr\{W_2 = 20\%\} = 0.01$	
0%	$p_{24} = \Pr\{W_2 = 0\%\} = 0.01$	

Assume that the performance effectiveness of each unit can be improved by using loaded redundancy. Let us suppose that at each moment of time, the performance effectiveness of a redundant group is equal to the level of performance of the best component of the redundant group. Thus, the behavior of Unit 1, consisting of the main component and single redundant element, can be depicted as shown in Figure 11.1. For Unit 2, the analogous process is presented in Figure 11.2.

Further, assume that the entire system (series connection of Unit 1 and Unit 2) is characterized by the worst level of effectiveness of its units at each moment of time. In Figure 11.3, one can see the system behavior for the case when both units consist of a single main element.

The problem is to find the optimal number of redundant elements in the series system described above:

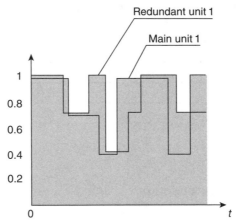

FIGURE 11.1 A realization of stochastic behavior of Unit 1, consisting of two elements, main and redundant. The shadowed area denotes the behavior of Unit 1.

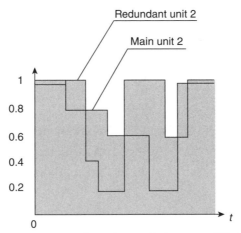

FIGURE 11.2 A realization of stochastic behavior of Unit 2, consisting of two elements, main and redundant. The shadowed area denotes the behavior of Unit 2.

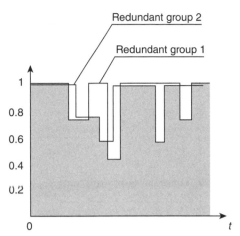

FIGURE 11.3 A realization of stochastic behavior of the entire system when both its units consist of a single main element. The shadowed area denotes the behavior of the system.

(1) Direct problem: Find such an allocation of redundant elements that delivers an average level of system performance greater than or equal to the required level of performance with a minimum possible cost of redundant elements.

(2) Inverse problem: Find such an allocation of redundant elements that delivers the maximum possible level of system performance under the condition that the total expenditures on redundant elements do not exceed the given total cost of redundant units.

Now consider the construction of a dominating sequence during the optimization process. In principle, one has to construct a table like Table 11.3 and choose members of the dominating sequence.

As one can see, in this case we deal with quadruplets of type:

```
{[Vector of numbers of redundant units];
 [Discrete distribution of performance levels];
 [Performance levels];
 [System cost]}.
```

TABLE 11.3 Construction of Dominating Sequence

		Number of redundant elements for Unit 1			
		0	1	2	. . .
Number of redundant elements for Unit 2	0	$X = (0, 0)$ P(0, 0) W(0, 0) C(0, 0)	$X = (1, 0)$ P(1, 0) W(1, 0) C(1, 0)	$X = (2, 0)$ P(2, 0) W(2, 0) C(2, 0)	. . .
	1	$X = (0, 1)$ P(0, 1) W(0, 1) C(0, 1)	$X = (1, 1)$ P(1, 1) W(1, 1) C(1, 1)	$X = (2, 1)$ P(2, 1) W(2, 1) C(2, 1)	. . .
	2	$X = (0, 2)$ P(0, 2) W(0, 2) C(0, 2)	$X = (1, 2)$ P(1, 2) W(1, 2) C(1, 2)	$X = (2, 2)$ P(2, 1) W(2, 2) C(2, 2)	. . .
	. . .	. . .	. . .	. . .	. . .

The problem complicates due to necessity of calculations because "probabilities of performance levels" and "performance levels" are not numbers but vectors that need a special type of calculation. This aspect will be demonstrated shortly. Here we would like to note that there is no necessity to calculate quadruplets for all cells of Table 11.1. Fortunately, we can use Kettelle's Algorithm: members of dominating sequences are located around the table's diagonal and the corresponding cells form a simple connected area. It allows the use of the "dichotomy tree" procedure, that is, avoiding unnecessary calculations by cutting non-perspective branches (see Fig. 11.4). Indeed, consider bordering cells around the simple connected area (they are marked as "x"). There are no dominating cells in the area located in the upper right, and there are no dominating cells in the area located in the lower left.

Thus, in this case, the calculations occur to be sufficiently compact. However, as we mentioned above, some special calculations for each redundant group have to be done.

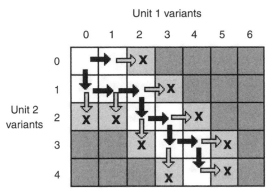

FIGURE 11.4 Example of excluding non-perspective branches. Black arrows are members of dominating sequence; gray ones are trial tests that led to non-perspective variants marked by "x." All cells marked with dark gray cannot contain dominating quadruplets.

In accordance with the calculating procedure described above, one has to consider the first variant $(0, 0)$, that is, just Unit 1 and Unit 2 with no redundancy at all, and find the quadruple. In this case, the resulting solution will be:

$$\{[0]; [(p_{11}, W_{11}), (p_{12}, W_{12}), (p_{13}, W_{13}), (p_{14}, W_{14})]; [c]_1\}$$
$$\otimes \{[0]; [(p_{21}, W_{21}), (p_{22}, W_{22}), (p_{23}, W_{23}), (p_{24}, W_{24})]; [c_2]\}$$
$$= \{0 \underset{\rightarrow}{\otimes} 0; [(p_{11}, W_{11}), (p_{12}, W_{12}), (p_{13}, W_{13}), (p_{14}, W_{14})]\} \tag{11.1}$$
$$\underset{UGF}{\otimes} \{0; [(p_{21}, W_{21}), (p_{22}, W_{22}), (p_{23}, W_{23}), (p_{24}, W_{24})]; c_1 \underset{+}{\otimes} c_2\}.$$

Here we use the following operators:

$\underset{\rightarrow}{\otimes}$ is an operator of forming a vector, that is, $j \otimes k = (j, k)$;

$\underset{UGF}{\otimes}$ is an operator equivalent to the U-function, that is,

$$\left(\sum_{j \in A} p_j z^{W_j}\right) \underset{UGF}{\otimes} \left(\sum_{k \in B} p_k z^{W_k}\right) = \sum_{\forall j, \forall k} p_j \cdot p_k z^{W_j \underset{min}{\otimes} W_k}, \quad \text{where, in turn,}$$

$W_j \underset{min}{\otimes} W_k = \min(W_j, W_k);$

$c_1 \underset{+}{\otimes} c_2$ is operator of summation, that is, $c_1 \underset{+}{\otimes} c_2 = c_1 + c_2.$

Of course, the same procedure can be presented in terms of U functions. One can write two "polynomials" of type

$$
\left(\sum_{j \in A} p_{1j} x^0 y^{W_{1j}} z^{c_{1j}} \right) \underset{UGF}{\otimes} \left(\sum_{k \in B} p_{2k} x^0 y^{W_{2j}} z^{c_{2j}} \right)
$$

$$
= \sum_{\forall j, \forall k} p_j \cdot p_k x^{\underset{min}{0 \otimes 0}} y^{\underset{min}{W_j \otimes W_k}} z^{\underset{+}{c_{2j} \otimes c_{2k}}} = \sum_{\forall j, \forall k} p_j \cdot p_k x^{(0,0)} y^{\max(W_{1j}, W_{2k})} z^{\underset{+}{c_{2j} + c_{2k}}}.
$$

$$(11.2)$$

Of course, the power of argument x has a very conditional sense: any value in "power" of vector has no common sense. To avoid such confusion, we will operate with sequences of triplets, quadruplets, and other "multiplets."

Let us continue the numerical example because it helps us to not explain relatively simple procedures on an unnecessarily formal level. We now return to the series system, consisting of two units without redundancy. Numerical results are presented in Table 11.4.

This leads to the following final result (see cells with the same background colors):

$$P^{(0.0)}(W_{syst} = 100\%) = 0.72;$$

$$P^{(0.0)}(W_{syst} = 80\%) = 0.171;$$

$$P^{(0.0)}(W_{syst} = 70\%) = 0.04 + 0.0095 = 0.0495;$$

$$P^{(0.0)}(W_{syst} = 40\%) = 0.032 + 0.0076 = 0.0396;$$

$$P^{(0.0)}(W_{syst} = 20\%) = 0.009 + 0.0005 + 0.004 = 0.0099;$$

$$P^{(0.0)}(W_{syst} = 0\%) = 0.008 + 0.0019 + 0.0001 + 0.0001 + 0.0009 + 0.0005 + 0.0004 = 0.0201.$$

Cost of additional units in this case equals 0. As one can easily calculate, the average level of the system performance is equal to

TABLE 11.4 Initial State of the Process of Optimization

(0, 0)	Unit 2			
	$p_{21} = 0.8$ $W_{21}^{(0)} = 100\%$	$p_{22} = 0.18$ $W_{22}^{(0)} = 80\%$	$p_{23} = 0.01$ $W_{23}^{(0)} = 20\%$	$p_{24} = 0.01$ $W_{24}^{(0)} = 0\%$
Unit 1 $p_{11} = 0.9$ $W_{11}^{(0)} = 100\%$	$p_{21} \cdot p_{11} = 0.72$ $\min(W_{21}^{(0)}, W_{11}^{(0)}) = 100\%$	$p_{22} \cdot p_{11} = 0.171$ $\min(W_{22}^{(0)}, W_{11}^{(0)}) = 80\%$	$p_{23} \cdot p_{14} = 0.009$ $\min(W_{23}^{(0)}, W_{11}^{(0)}) = 20\%$	$p_{24} \cdot p_{14} = 0.009$ $\min(W_{21}^{(0)}, W_{11}^{(0)}) = 0\%$
$p_{12} = 0.05$ $W_{12}^{(0)} = 70\%$	$p_{21} \cdot p_{12} = 0.04$ $\min(W_{21}^{(0)}, W_{12}^{(0)}) = 70\%$	$p_{22} p_{12} = 0.0095$ $\min(W_{22}^{(0)}, W_{12}^{(0)}) = 70\%$	$p_{23} \cdot p_{14} = 0.0005$ $\min(W_{23}^{(0)}, W_{12}^{(0)}) = 20\%$	$p_{24} \cdot p_{14} = 0.0005$ $\min(W_{21}^{(0)}, W_{11}^{(0)}) = 0\%$
$p_{13} = 0.04$ $W_{13}^{(0)} = 40\%$	$p_{21} \cdot p_{13} = 0.032$ $\min(W_{21}^{(0)}, W_{13}^{(0)}) = 40\%$	$p_{22} p_{13} = 0.0076$ $\min(W_{22}^{(0)}, W_{13}^{(0)}) = 40\%$	$p_{23} \cdot p_{14} = 0.0004$ $\min(W_{23}^{(0)}, W_{13}^{(0)}) = 20\%$	$p_{24} \cdot p_{14} = 0.0004$ $\min(W_{21}^{(0)}, W_{11}^{(0)}) = 0\%$
$p_{14} = 0.01$ $W_{14}^{(0)} = 0\%$	$p_{21} \cdot p_{14} = 0.008$ $\min(W_{21}^{(0)}, W_{14}^{(0)}) = 0\%$	$p_{22} \cdot p_{14} = 0.0019$ $\min(W_{22}^{(0)}, W_{14}^{(0)}) = 0\%$	$p_{23} \cdot p_{14} = 0.0001$ $\min(W_{23}^{(0)}, W_{14}^{(0)}) = 0\%$	$p_{24} \cdot p_{14} = 0.0001$ $\min(W_{23}^{(0)}, W_{14}^{(0)}) = 0\%$

TABLE 11.5 Forehand Calculation of Performance Levels Distribution for Unit 1, Consisting of Two Elements, Main and Redundant

Element 1	Element 1			
	$p_{11} = 0.9$ $W_{11}^{(0)} = 100\%$	$p_{12} = 0.05$ $W_{12}^{(0)} = 70\%$	$p_{13} = 0.04$ $W_{13}^{(0)} = 40\%$	$p_{14} = 0.01$ $W_{14}^{(0)} = 0\%$
$p_{11} = 0.9$ $W_{11}^{(0)} = 100\%$	$(p_{11})^2 = 0.81$ $W_{11}^{(0)} = 100\%$	$p_{12} \cdot p_{11} = 0.045$ $\max(W_{12}^{(0)}, W_{11}^{(0)}) = 100\%$	$p_{13} \cdot p_{11} = 0.036$ $\max(W_{13}^{(0)}, W_{11}^{(0)}) = 100\%$	$p_{14} \cdot p_{11} = 0.009$ $\max(W_{14}^{(0)}, W_{11}^{(0)}) = 100\%$
$p_{12} = 0.05$ $W_{12}^{(0)} = 70\%$	$p_{11} \cdot p_{12} = 0.045$ $\max(W_{11}^{(0)}, W_{12}^{(0)}) = 100\%$	$(p_{12})^2 = 0.025$ $\max(W_{12}^{(0)}, W_{12}^{(0)}) = 70\%$	$p_{13} \cdot p_{12} = 0.002$ $\max(W_{13}^{(0)}, W_{12}^{(0)}) = 70\%$	$p_{14} \cdot p_{12} = 0.0005$ $\max(W_{14}^{(0)}, W_{12}^{(0)}) = 70\%$
$p_{13} = 0.04$ $W_{32}^{(0)} = 40\%$	$p_{11} \cdot p_{13} = 0.036$ $\max(W_{11}^{(0)}, W_{32}^{(0)}) = 100\%$	$p_{12} \cdot p_{13} = 0.002$ $\max(W_{12}^{(0)}, W_{32}^{(0)}) = 70\%$	$(p_{13})^2 = 0.0016$ $W_{32}^{(0)} = 40\%$	$p_{14} \cdot p_{13} = 0.0004$ $\max(W_{14}^{(0)}, W_{32}^{(0)}) = 40\%$
$p_{14} = 0.01$ $W_{14}^{(0)} = 0\%$	$p_{11} \cdot p_{14} = 0.009$ $\max(W_{11}^{(0)}, W_{14}^{(0)}) = 100\%$	$p_{12} \cdot p_{14} = 0.0005$ $\max(W_{12}^{(0)}, W_{14}^{(0)}) = 70\%$	$p_{13} \cdot p_{14} = 0.0004$ $\max(W_{13}^{(0)}, W_{14}^{(0)}) = 40\%$	$(p_{14})^2 = 0.0001$ $W_{14}^{(0)} = 0\%$

$$W_{syst}^{(0,0)} = 0.72 + 0.171 \cdot 0.8 + 0.0497 \cdot 0.7 + 0.0396 \cdot 0.5 + 0.0095 \cdot 0.2$$
$$\approx 0.9092.$$

Now let's make trial steps to the neighbor cells: check cells (1, 0) and (0, 1). Let us start with cell (1, 0) as it shown in Figure 11.4. First, find the distribution of performance levels for Unit 1 consisting of two elements, main and redundant.

On the basis of this table, one gets for Unit 1 the following distribution

$$\Pr\{W_1^{(1)} = 100\%\} = P_{11}^{(1)} = (p_{11})^2 + 2p_{11} \cdot (p_{12} + p_{13} + p_{14})$$
$$= 0.81 + 2 \cdot (0.045 + 0.036 + 0.009) = 0.99;$$

$$\Pr\{W_1^{(1)} = 70\%\} = P_{12}^{(1)} = (p_{12})^2 + 2 \cdot p_{12} \cdot (p_{13} + p_{14})$$
$$= 0.025 + 2 \cdot 0.025(0.002 + 0.0005) = 0.0075;$$

$$\Pr\{W_1^{(1)} = 40\%\} = P_{13}^{(1)} = (p_{13})^2 + 2p_{13} \cdot p_{14} = 0.0016 + 2 \cdot 0.0016 \cdot 0.0004$$
$$\approx 0.0016;$$

$$\Pr\{W_1^{(1)} = 0\%\} = P_{14}^{(1)} = (p_{14})^2 = 0.0001.$$

Let us assume that the first step is made from (0, 0) to (1, 0) as is presented in Figure 11.5.

Using the results, presented above, one can compile Table 11.6, which gives performance levels distribution for the system characterized by vector of redundant elements $X = (1, 0)$.

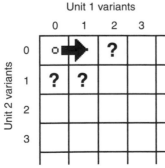

FIGURE 11.5 Direction of step 1 of the optimization process.

TABLE 11.6 Step 1 of the Optimization Process

$C_{system} = (1, 0)$	Unit 2			
	$p_{21} = 0.8$ $W_{21}^{(0)} = 100\%$	$p_{22} = 0.19$ $W_{22}^{(0)} = 80\%$	$p_{23} = 0.01$ $W_{23}^{(0)} = 20\%$	$p_{24} = 0.01$ $W_{24}^{(0)} = 0\%$
Unit 1 $P_{11}^{(1)} = 0.99$ $W_{11}^{(1)} = 100\%$	$p_{21} \cdot P_{11}^{(1)} = 0.792$ $\min(W_{21}^{(0)}, W_{11}^{(1)}) = 100\%$	$p_{22} \cdot P_{11}^{(1)} = 0.188$ $\min(W_{22}^{(0)}, W_{11}^{(1)}) = 80\%$	$p_{23} \cdot P_{11}^{(1)} \approx 0.01$ $\min(W_{23}^{(0)}, W_{11}^{(1)}) = 20\%$	$p_{24} \cdot P_{11}^{(1)} \approx 0.01$ $\min(W_{24}^{(0)}, W_{11}^{(1)}) = 0\%$
$P_{12}^{(1)} = 0.0075$ $W_{12}^{(1)} = 70\%$	$p_{21} \cdot P_{12}^{(1)} = 0.006$ $\min(W_{21}^{(0)}, W_{12}^{(1)}) = 70\%$	$p_{22} \cdot P_{12}^{(1)} \approx 0.0014$ $\min(W_{22}^{(0)}, W_{12}^{(1)}) = 70\%$	$p_{23} \cdot P_{12}^{(1)} \approx 0.0001$ $\min(W_{23}^{(0)}, W_{12}^{(1)}) = 20\%$	$p_{24} \cdot P_{12}^{(1)} = 0.0001$ $\min(W_{24}^{(0)}, W_{12}^{(1)}) = 0\%$
$P_{13}^{(1)} = 0.0016$ $W_{13}^{(1)} = 40\%$	$p_{21} \cdot P_{13}^{(1)} \approx 0.0013$ $\min(W_{21}^{(0)}, W_{13}^{(1)}) = 40\%$	$p_{22} \cdot P_{13}^{(1)} = 0.0003$ $\min(W_{22}^{(0)}, W_{13}^{(1)}) = 40\%$	$p_{23} \cdot P_{13}^{(1)} \approx 0$ $\min(W_{23}^{(0)}, W_{13}^{(1)}) = 20\%$	$p_{24} \cdot P_{13}^{(1)} \approx 0$ $\min(W_{24}^{(0)}, W_{13}^{(1)}) = 0\%$
$P_{14}^{(1)} = 0.0001$ $W_{14}^{(1)} = 0\%$	$p_{21} \cdot P_{14}^{(1)} = 0.0001$ $\min(W_{21}^{(0)}, W_{14}^{(1)}) = 0\%$	$p_{22} \cdot P_{14}^{(1)} \approx 0$ $\min(W_{22}^{(0)}, W_{14}^{(1)}) = 0\%$	$p_{23} \cdot P_{14}^{(1)} \approx 0$ $\min(W_{23}^{(0)}, W_{14}^{(1)}) = 0\%$	$p_{24} \cdot P_{14}^{(1)} \approx 0$ $\min(W_{24}^{(0)}, W_{14}^{(1)}) = 0\%$

This leads to the following final result:

$$P^{(1.0)}(W_{syst} = 100\%) = 0.792;$$

$$P^{(0.0)}(W_{syst} = 80\%) = 0.188;$$

$$P^{(0.0)}(W_{syst} = 70\%) = 0.006 + 0.0014 = 0.0074;$$

$$P^{(0.0)}(W_{syst} = 40\%) = 0.0013 + 0.0003 = 0.0016;$$

$$P^{(0.0)}(W_{syst} = 20\%) = 0.01 + 0.0001 = 0.0101;$$

$$P^{(0.0)}(W_{syst} = 0\%) = 0.008 + 0.0019 + 0.0001 + 0.0001 + 0.0009$$
$$+ 0.0005 + 0.0004 = 0.0201.$$

Cost of additional units in this case equals 1. Average system's performance level equals

$$W_{syst}^{(1,0)} = 0.792 + 0.188 \cdot 0.8 + 0.0497 \cdot 0.7 + 0.0396 \cdot 0.4 + 0.0095 \cdot 0.2$$
$$\approx 0.9502.$$

Then try another neighbor cell, namely (0, 1). Beforehand, one has to perform an additional calculation of performance levels distribution for Unit 2 consisting of two elements, main and redundant.

It is necessary to note that for parallel connection of multistate elements (that compiles a unit), one should realistically assume that the level of performance of the unit is equal to maximum among all currently operating elements. So, Table 11.7 represents results of calculation for Unit 2 that consists of two identical elements.

On the basis of this table, one gets for Unit 2, consisting of two elements, the following distribution:

$$\Pr\{W_2^{(1)} = 100\%\} = P_{21}^{(1)} = (p_{21})^2 + 2p_{21} \cdot (p_{22} + p_{23} + p_{34})$$
$$= 0.64 + 2 \cdot 0.8 \cdot (0.045 + 0.036 + 0.008) \approx 0.7709;$$

$$\Pr\{W_2^{(1)} = 80\%\} = P_{22}^{(1)} = (p_{22})^2 + 2 \cdot p_{22} \cdot (p_{23} + p_{24})$$
$$= 0.0361 + 2 \cdot 0.0361(0.0002 + 0.0002) \approx 0.0361;$$

TABLE 11.7 Forehand Calculation of Performance Level Distribution for Unit 2, Consisting of Two Elements, Main and Redundant

Ele-ment 2	Element 2			
	$p_{21} = 0.8$ $W_{21}^{(0)} = 100\%$	$p_{22} = 0.19$ $W_{22}^{(0)} = 80\%$	$p_{23} = 0.01$ $W_{23}^{(0)} = 20\%$	$p_{24} = 0.01$ $W_{24}^{(0)} = 0\%$
$p_{21} = 0.8$ $W_{21}^{(0)} = 100\%$	$(p_{21})^2 = 0.64$ $W_{21}^{(0)} = 100\%$	$p_{22} \cdot p_{21} = 0.045$ $\max(W_{22}^{(0)}, W_{21}^{(0)}) = 100\%$	$p_{23} \cdot p_{21} = 0.036$ $\max(W_{23}^{(0)}, W_{21}^{(0)}) = 100\%$	$p_{24} \cdot p_{21} = 0.008$ $\max(W_{24}^{(0)}, W_{21}^{(0)}) = 100\%$
$p_{22} = 0.19$ $W_{21}^{(0)} = 80\%$	$p_{21} \cdot p_{22} = 0.152$ $\max(W_{21}^{(0)}, W_{21}^{(0)}) = 100\%$	$(p_{22})^2 = 0.0361$ $W_{22}^{(0)} = 80\%$	$p_{23} \cdot p_{22} = 0.0002$ $\max(W_{23}^{(0)}, W_{21}^{(0)}) = 80\%$	$p_{24} \cdot p_{22} = 0.0002$ $\max(W_{24}^{(0)}, W_{21}^{(0)}) = 80\%$
$p_{23} = 0.01$ $W_{23}^{(0)} = 20\%$	$p_{21} \cdot p_{23} = 0.008$ $\max(W_{21}^{(0)}, W_{23}^{(0)}) = 100\%$	$p_{22} \cdot p_{23} = 0.0002$ $\max(W_{22}^{(0)}, W_{23}^{(0)}) = 80\%$	$(p_{23})^2 = 0.0001$ $W_{23}^{(0)} = 20\%$	$p_{24} \cdot p_{23} = 0.0001$ $\max(W_{24}^{(0)}, W_{23}^{(0)}) = 20\%$
$p_{24} = 0.01$ $W_{24}^{(0)} = 0\%$	$p_{21} \cdot p_{24} = 0.008$ $\max(W_{21}^{(0)}, W_{24}^{(0)}) = 100\%$	$p_{22} \cdot p_{24} = 0.0002$ $\max(W_{22}^{(0)}, W_{24}^{(0)}) = 70\%$	$p_{23} \cdot p_{24} = 0.0001$ $\max(W_{23}^{(0)}, W_{24}^{(0)}) = 20\%$	$(p_{24})^2 = 0.0001$ $W_{24}^{(0)} = 0\%$

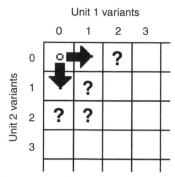

FIGURE 11.6 Direction of step 2 of the optimization process.

$$\Pr\{W_2^{(1)} = 20\%\} = P_{23}^{(1)} = (p_{13})^2 + 2p_{13} \cdot p_{14} = 0.0001 + 0.0001 + 0.0001$$
$$= 0.0003;$$

$$\Pr\{W_2^{(1)} = 0\%\} = P_{24}^{(1)} = (p_{14})^2 = 0.0001.$$

After such preparations, one can go to step 2 (see Fig. 11.6).

This step consists of construction of Table 11.8 and presents the system's performance levels distribution for the system configuration characterized by a vector of redundant elements $X = (0, 1)$.

This leads to the following final result:

$$P^{(1.0)}(W_{syst} = 100\%) = 0.6038;$$

$$P^{(0.0)}(W_{syst} = 80\%) = 0.0325;$$

$$P^{(0.0)}(W_{syst} = 70\%) = 0.0386 + 0.0018 = 0.0404;$$

$$P^{(0.0)}(W_{systt} = 40\%) = 0.0308 + 0.0014 = 0.0322;$$

$$P^{(0.0)}(W_{systt} = 20\%) \approx 0.0003;$$

$$P^{(0.0)}(W_{syst} = 0\%) = 0.0077 + 0.0004 + 0.0001 \approx 0.0082.$$

Cost of additional units in this case equals 2 units of cost. The average system's performance level equals

TABLE 11.8 Step 2 of the Process of Optimization

$C_{system} = (0, 1)$		Unit 2			
		$P_{21}^{(1)} = 0.7709$ $W_{21}^{(1)} = 100\%$	$P_{22}^{(1)} = 0.0361$ $W_{22}^{(1)} = 80\%$	$P_{23}^{(1)} = 0.0003$ $W_{23}^{(1)} = 20\%$	$P_{24}^{(1)} = 0.0001$ $W_{24}^{(1)} = 0\%$
Unit 1	$p_{11} = 0.9$ $W_{11}^{(0)} = 100\%$	$P_{21}^{(1)} \cdot p_{11} \approx 0.6038$ $\min(W_{21}^{(1)}, W_{11}^{(0)}) = 100\%$	$P_{22}^{(1)} \cdot p_{11} \approx 0.0325$ $\min(W_{22}^{(1)}, W_{11}^{(1)}) = 80\%$	$P_{23}^{(1)} \cdot p_{11} \approx 0.0003$ $\min(W_{23}^{(1)}, W_{11}^{(1)}) = 20\%$	$P_{24}^{(1)} \cdot p_{11} \approx 0.0001$ $\min(W_{24}^{(1)}, W_{11}^{(1)}) = 0\%$
	$p_{12} = 0.05$ $W_{12}^{(0)} = 70\%$	$P_{21}^{(1)} \cdot p_{12} \approx 0.0386$ $\min(W_{21}^{(0)}, W_{12}^{(1)}) = 70\%$	$P_{22}^{(1)} \cdot p_{12} \approx 0.0018$ $\min(W_{22}^{(0)}, W_{12}^{(1)}) = 70\%$	$P_{23}^{(1)} \cdot p_{12} \approx 0$ $\min(W_{23}^{(0)}, W_{12}^{(1)}) = 20\%$	$P_{24}^{(1)} \cdot p_{12} \approx 0$ $\min(W_{24}^{(0)}, W_{12}^{(1)}) = 0\%$
	$p_{13} = 0.04$ $W_{32}^{(0)} = 40\%$	$P_{21}^{(1)} \cdot p_{13} \approx 0.0308$ $\min(W_{21}^{(0)}, W_{13}^{(1)}) = 40\%$	$P_{22}^{(1)} \cdot p_{13} \approx 0.0014$ $\min(W_{22}^{(0)}, W_{13}^{(1)}) = 40\%$	$P_{23}^{(1)} \cdot p_{13} \approx 0$ $\min(W_{23}^{(0)}, W_{13}^{(1)}) = 20\%$	$P_{24}^{(1)} \cdot p_{13} \approx 0$ $\min(W_{24}^{(0)}, W_{13}^{(1)}) = 0\%$
	$p_{14} = 0.01$ $W_{14}^{(0)} = 0\%$	$P_{21}^{(1)} \cdot p_{14} \approx 0.00771$ $\min(W_{21}^{(0)}, W_{14}^{(1)}) = 0\%$	$P_{22}^{(1)} \cdot p_{14} \approx 0.0004$ $\min(W_{22}^{(0)}, W_{14}^{(1)}) = 0\%$	$P_{23}^{(1)} \cdot p_{14} \approx 0$ $\min(W_{23}^{(0)}, W_{14}^{(1)}) = 0\%$	$P_{24}^{(1)} \cdot p_{14} \approx 0$ $\min(W_{24}^{(0)}, W_{14}^{(1)}) = 0\%$

$$W_{syst}^{(0,1)} = 0.6038 + 0.0325 \cdot 0.8 + 0.0404 \cdot 0.7 + 0.0322 \cdot 0.4$$
$$+ 0.0003 \cdot 0.2 \approx 0.671.$$

Thus, for vector $(1, 0)$, one has additional cost equal to 1 and $W_{syst}^{(1,0)} \approx 0.9502$ and for vector $(0, 1)$ corresponding values equal to 2 and 0.671, so system configuration $(1, 0)$ is dominating over configuration $(0, 1)$, since a higher average performance level delivers with fewer expenses. It means that all vectors of type $(0, k)$ are excluded from further analysis.

The next cells for which current trials have to be done are $(1, 1)$ and $(2, 0)$.

Avoiding simple but cumbersome calculations, let us present only the final results (see Table 11.9).

Table 11.9 is constructed as is shown in Figure 11.7. Table 11.9 probably needs some explanation. Vector $(2, 0)$ is dominated by vector $(0, 1)$, since vector $(0, 1)$ is characterized by higher performance level for the same total cost of redundant units. So, all vectors of type $(3, 0)$, $(4, 0)$, . . ., $(k, 0)$, . . . are excluded from further consideration. One observes the same type of domination for the following pairs: $(3, 1)$ is dominated by $(1, 2)$, vector $(0, 2)$ is dominated by $(1, 1)$, vector $(1, 3)$ is dominated by $(2, 2)$, and so on.

Such trials and the selection of dominating vectors continued until the appearance of the first vector with the average level of performance higher than the required value of W^0 for the direct problem of optimal redundancy, or until the total expense of all redundant elements does not exceed the given value C^0 for the inverse problem. These comments become absolutely transparent if one takes a look at Figure 11.8.

From Table 11.9, one can see that the optimal solution for the requirement that the average level of system performance is not less than $W^0 = 0.999$ is delivered by vector $(3, 2)$, and the total expenses of redundant elements are 7 cost units. For the total expenses of redundant elements limited by $C^0 \leq 5$ cost units, one

TABLE 11.9 Costs and Levels of Performance for Different Vectors of Redundant Elements

		Unit 1: Number of redundant elements						
		0	1	2	3	4	5	6
Unit 2: Number of redundant elements	0	C = 0 W = 90.16	C = 1 W = 94.26	C = 2 W = 94.57				…
	1	C = 2 W = 94.68	C = 3 W = 99.16	C = 4 W = 99.50	C = 5 W = 99.53			…
	2	C = 4 W = 95.03	C = 5 W = 99.54	C = 6 W = 99.89	C = 7 W = 99.92	?		…
	3		C = 7 W = 99.61	C = 8 W = 99.95	?			…
	4			?	…			…
	…	…	…	…			…	…

Note: Light gray color indicates dominated cells; dark gray color, non-prospective variants.

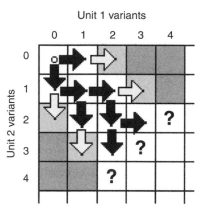

FIGURE 11.7 The process of step-by-step development of the optimization process.

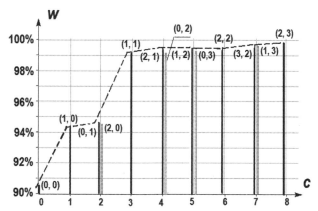

FIGURE 11.8 Depiction of the process of compiling the dominating sequence.

gets the maximum possible solution as vector (1, 2) that is characterized by $W = 99.54\%$.

It is interesting to see what happens with the optimal solution if one changes the costs of elements. Let us assume that for the same system, cost of a single redundant element of the first type is $c_1 = 2$ and the cost of an element of the second type $c_2 = 1$.

In this case, optimal solutions found in Table 11.10 are: For the direct problem vector (2, 3), for which $W = 99.95\%$ and total expenses

TABLE 11.10 Costs and Levels of Performance for Different Vectors of Redundant Elements for New Element's Costs

Unit 2: Number of redundant elements	Unit 1: Number of redundant elements						
	0	1	2	3	4	5	6
0	$C = 0$ $W = 90.156$	$C = 2$ $W = 94.26456$					...
1	$C = 1$ $W = 94.68072$	$C = 3$ $W = 99.16318$	$C = 5$ $W = 99.50462$				...
2	$C = 2$ $W = 95.02683$	$C = 4$ $W = 99.54058$	$C = 6$ $W = 99.88507$	$C = 8$ $W = 99.9156$			...
3	$C = 3$ $W = 95.08558$	$C = 5$ $W = 99.60514$	$C = 7$ $W = 99.9502$	?	...		
4		$C = 6$ $W = 99.61784$	?		...	...	
...	...	...		...			...

Note: Light gray color indicates dominated cells; dark gray color, non-prospective variants.

on redundant elements are equal to 7 cost units, and for the inverse problem the solution is (1, 3), for which $W = 99.54\%$ and total expenses are $C = 5$.

Solution of optimal redundancy problems for a system consisting of several multilevel units seems a bit cumbersome. However, let us note that all enumerative methods, like dynamic programming, are practically unsolvable without computerizing calculations. The numerical example above was solved with the help of simple programs using Microsoft Excel.

For complex systems consisting of n multiple multistate units, one can compile a simple program for a mainframe computer. The algorithm should include the following steps.

1. Take an n-dimensional vector of redundant elements $X^{(0)} = (x_1^{(0)} = 0, x_2^{(0)} = 0, \dots, x_n^{(0)} = 0)$.
2. Perform calculations to get initial pair of values $(W_{syst}^{(0)}, C_{syst}^{(0)})$ (see Table 11.2).
3. Put calculated pair $(W_{syst}^{(0)}, C_{syst}^{(0)})$ into a list of dominating solutions,
4. Generate vectors $X_i^{(1)}$ such that each of them distinguishes from $X^{(0)}$ by changing number of elements of Unit I on one, that is, $X_i^{(1)} = (x_1^{(0)} = 0, x_2^{(0)} = 0, \dots, x_i^{(0)} = 1, \dots, x_n^{(0)} = 0)$.
5. For each $X_i^{(1)}, \overline{i = 1, n}$ calculate new values of $P_{ik_i}^{(1)}$, for all k_i where k_i is the number of performance levels of Unit I.
6. Perform calculations to get n pairs $(W_1^{(1)}, C_1^{(1)}), (W_2^{(1)}, C_2^{(1)}), \dots,$ $(W_n^{(1)}, C_n^{(1)})$ for all vectors.

Such a solution appears a bit clumsy and laborious. However, the computer calculating program is relatively simple and the solution can be obtained easily enough; final results are presented in the form of a dominating sequence (in Kettelle's terminology), so the solutions for direct and/or inverse optimal redundancy problems can be easily found.

We restrict ourselves by consideration of this simple and more or less transparent illustrative example. In the last few years this problem has generated a number of interesting and theoretically deep publications, as the reader can see from the Bibliography. However, we think that more detailed consideration of this problem could lead us too far from the "highway" of main practical optimal redundancy tasks.

CHRONOLOGICAL BIBLIOGRAPHY

Reinschke, K. 1985. "Systems consisting of units with multiple states." In *Handbook: Reliability of Technical Systems*, I. Ushakov, ed. Sovetskoe Radio.

El-Neweihi, E., Proschan, F., and Sethuraman, J. 1988. "Optimal allocation of multistate components." In *Handbook of Statistics*, vol. 7: *Quality Control and Reliability*, P. R. Krishnaiah and C. R. Rao, eds. North-Holland.

Ushakov, I. 1988. "Reliability analysis of multi-state systems by means of modified generating function." *Elektronische Informationsverarbeitung und Kybernetik*, no. 3.

Levitin, G., and Lisnianski, A. 1998. "Joint redundancy and maintenance optimization for multistate series-parallel systems." *Reliability Engineering and System Safety*, no. 64,.

Levitin, G., Lisnianski, A., Ben Haim, H., and Elmakis, D. 1998. "Redundancy optimization for series-parallel multi-state systems." *IEEE Transactions on Reliability*, no. 2.

Levitin, G., and Lisnianski, A. 1999. "Importance and sensitivity analysis of multistate systems using universal generating functions method." *Reliability Engineering and System Safety*, no. 65.

Levitin, G., and Lisnianski, A. 2000. "Optimal replacement scheduling in multistate series-parallel systems." *Quality and Reliability Engineering International*, no. 16.

Levitin, G., Lisnianski, A., and Ben Haim, H. 2000. "Structure optimization of multi-state system with time redundancy." *Reliability Engineering and System Safety*, no. 67.

Levitin, G. 2001. "Redundancy optimization for multi-state system with fixed resource-requirements and unreliable sources." *IEEE Transactions on Reliability*, no. 50.

Levitin, G., and Lisnianski, A. 2001. "A new approach to solving problems of multi-state system reliability optimization." *Quality and Reliability Engineering International*, no. 47.

Levitin, G., and Lisnianski, A. 2001. "Structure optimization of multi-state system with two failure modes." *Reliability Engineering and System Safety*, no. 72.

Levitin, G. 2002. "Optimal allocation of multi-state elements in linear consecutively-connected systems with delays." *International Journal of Reliability Quality and Safety Engineering*, no. 9.

Levitin, G., Lisnianski, A., and Ushakov, I. 2002. "Multi-state system reliability: from theory to practice." *Proceedings of the 3rd Internationall Conference on Math ematical Models in Reliability*, Trondheim, Norway.

Levitin, G. 2003. "Optimal allocation of multi-state elements in linear consecutively-connected systems." *IEEE Transactions on Reliability*, no. 2.

Levitin, G. 2003. "Optimal multilevel protection in series-parallel systems." *Reliability Engineering and System Safety*, no. 81.

Levitin, G., Dai, Y., Xie, M., and Poh, K. L. 2003. "Optimizing survivability of multi-state systems with multi-level protection by multi-processor genetic algorithm." *Reliability Engineering and System Safety*, no. 82.

Levitin, G., Lisnianski, A., and Ushakov, I. 2003. "Reliability of multi-state systems: a historical overview." *Mathematical and Statistical Methods in Reliability*, no. 7.

Lisnianski, A., and Levitin, G. 2003. *Multi-State System Reliability: Assessment, Optimization and Applications*. World Scientific.

Levitin, G. 2004. "A universal generating function approach for analysis of multi-state systems with dependent elements." *Reliability Engineering and System Safety*, no. 3.

Nourelfath, M., and Dutuit, Y. 2004. "A combined approach to solve the redundancy optimization problem for multi-state systems under repair policies." *Reliability Engineering and System Safety*, no. 3.

Ramirez-Marquez, J. E., and Coit, D. W. 2004. "A heuristic for solving the redundancy allocation problem for multi-state series-parallel systems." *Reliability Engineering and System Safety*, no. 83.

Ding, Y., and Lisnianski, A. 2008. "Fuzzy universal generating functions for multi-state system reliability assessment." *Fuzzy Sets and Systems*, no. 3.

Tian, Z, Zuo, M. J., and Huang, H. 2008. "Reliability-redundancy allocation for multi-state series-parallel systems." *IEEE Transactions on Reliability*, no. 2.

CHAPTER *12*

CASE STUDIES

12.1 SPARE SUPPLY SYSTEM FOR WORLDWIDE TELECOMMUNICATION SYSTEM GLOBALSTAR

The software tool Optimal Spare Allocator (OSA) has been designed at QUALCOMM (San Diego, California) by the International Group on Reliability (IGOR). The International Group on Reliability includes Igor Ushakov (the lead), Dr. Sergei Antonov, Dr. Mikhail Konovalov, Dr. Sergei Shibanov, Dr. Sergei Shorgin, and Dr. Igor Sokolov (all from the Russian Academy of Sciences). OSA has been developed for a spare supply system of the worldwide low orbit satellite telecommunication system Globalstar. The OSA computer model allows for solving direct and inverse problems of optimal redundancy for hierarchical structure of spare stocks. It has a user-friendly interface and a convenient reporting utility.

Optimal Resource Allocation: With Practical Statistical Applications and Theory,
First Edition. Igor A. Ushakov.
© 2013 John Wiley & Sons, Inc. Published 2013 by John Wiley & Sons, Inc.

12.1.1 General Description of the Spare Support System

Satellite telecommunication system Globalstar has a number of ground base stations (gateways) dispersed all over the world. Successful operation of such a complex system depends on the ability to perform fast restoration of its operational ability after a failure. It can be reached by designing a maintenance system with hierarchical spare supply stocks of field replaceable units (FRU).

This supply system has a hierarchical structure with three levels: central spare stock (CSS), regional spare stocks (RSS), and on-site spare stocks (OSS) (see Fig. 12.1).

Diversity of gateways and installing new ones leads to permanent change of FRU requests for replacements. Such a situation leads to the necessity of designing a computer tool that is both powerful enough and easily operational enough to solve the problem of optimal spare allocation.

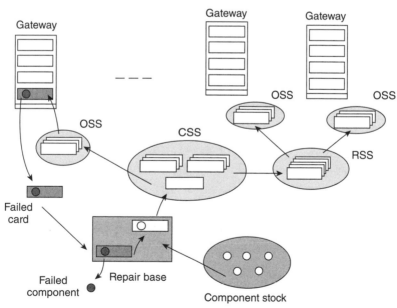

FIGURE 12.1 A hierarchical spare supply system.

The problems that arise are: (1) finding an optimal allocation of spare FRUs at each OSS depending on the size of the corresponding gateway, (2) finding the geographical location and size of each RSS and finding an optimal allocation of spare FRUs, and (3) finding the size of the CSS with optimal allocation of spare FRUs.

Gateway equipment consists of replaceable units. After each failure, a corresponding FRU from the OSS replaces the failed unit. A failed unit is sent to the repair base and after each failure a request is sent to the RSS, and the needed FRU is sent without delay (the so-called advance delivery). Regional and central stocks are usually supplied periodically: a request for replenishment is sent as soon as the level of stock declines to some critical level.

12.1.2   OSS, RSS, and CSS

We assume that gateways are highly reliable and their units are independent, so we neglect the possibility of overlapping of system down times due to different causes. For highly reliable systems, the approximate formula for the OSS unreliability coefficient, Q_{OSS}, is

$$Q_{OSS} \approx \sum_{1 \le k \le N} \beta_k q_k(x_k). \tag{12.1}$$

The weights in Equation (12.1) are defined as $\beta_k = \lambda_k n_k \left(\sum_{1 \le k \le N} \lambda_k n_k \right)^{-1}$, $q_k(x_k)$ = unreliability coefficient of units of type k (cumulative Poisson function with parameter $a_k = n_k \lambda_k \theta$), and x_k = number of spares of type k in the OSS. For highly reliable systems, the approximate formula for the OSS unavailability coefficient, U_{OSS}, is

$$U_{OSS} \approx \theta \sum_{k=1}^{N} \frac{\lambda_k n_k q_k(x_k)}{x_k + 1}, \tag{12.2}$$

where θ is the time delay corresponding to advance delivery.

A regional spare stock is periodically replenished from the central spare stock. The number and location of gateways, which are served by a particular RSS, may change in time with the development of the telecommunication system. It seems that the best index characterizing the RSS is its reliability coefficient. The same might be said about the CSS, which is replenished by production (probably with a different period for different types of units). In principle, the solution for these cases is similar to the previous one with the difference that the advance delivery period starts with the installation of a failed unit.

12.1.3 Software Tool OSA (Optimal Spare Allocator)

OSA is a GUI-driven, user-friendly tool designed to solve the direct and inverse problems of optimal redundancy. Due to its terrestrial dispersion, the spare supply system has a multi-level hierarchical structure, as shown in Figure 12.2. In OSA it is possible to use a detailed branching type scheme with all gateways' stocks shown, as in Figure 12.3.

For solving the optimal redundancy problem, the OSA tool uses the steepest descent method with insignificant modifications. The

FIGURE 12.2 OSA tool: map of hierarchical stock system.

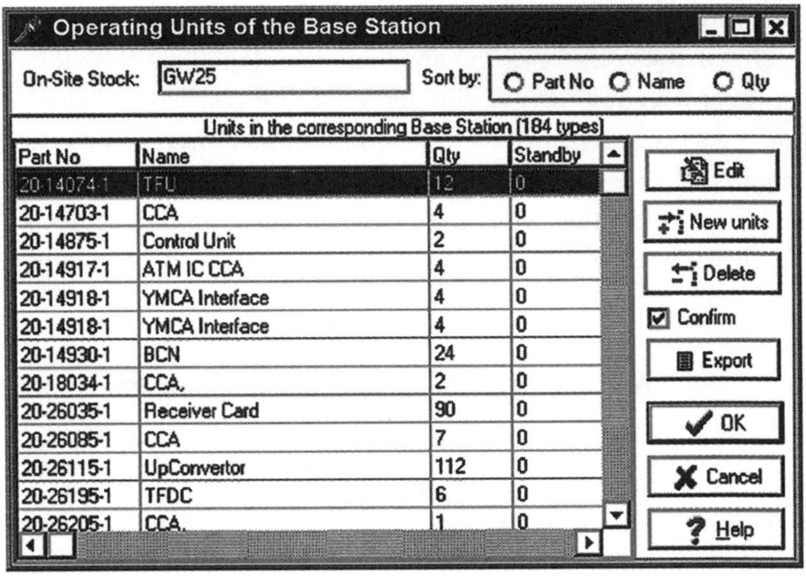

Operating Units of the Base Station					
On-Site Stock: GW25			Sort by: ○ Part No ○ Name ○ Qty		
Units in the corresponding Base Station (184 types)					
Part No	Name		Qty	Standby	
20-14074-1	TFU		12	0	
20-14703-1	CCA		4	0	
20-14875-1	Control Unit		2	0	
20-14917-1	ATM IC CCA		4	0	
20-14918-1	YMCA Interface		4	0	
20-14918-1	YMCA Interface		4	0	
20-14930-1	BCN		24	0	
20-18034-1	CCA,		2	0	
20-26035-1	Receiver Card		90	0	
20-26085-1	CCA		7	0	
20-26115-1	UpConvertor		112	0	
20-26195-1	TFDC		6	0	
20-26205-1	CCA,		1	0	

Buttons: Edit, New units, Delete, Confirm, Export, OK, Cancel, Help

FIGURE 12.3 OSA tool: a hierarchical spare supply system structure.

program's main window includes a menu of all available commands and a toolbar with the most frequently used operations.

The "Parameters" window (see Fig. 12.4) allows the user to choose a goal function: probability of stock failure (no available FRU of needed type at a failure moment). It suggests method of calculation and gives a possibility to choose what kind of optimal redundancy problem is to be solved. The user also enters the replenishment policy (since the computer program is the same for all levels of stocks).

Other options are rather specialized and we will not describe them.

The window shown in Figure 12.5 presents a list of units with their main parameters needed for solving the optimal allocation problem.

The next window (Fig. 12.6) shows specifications of the system: how many units of each type are installed within the system and

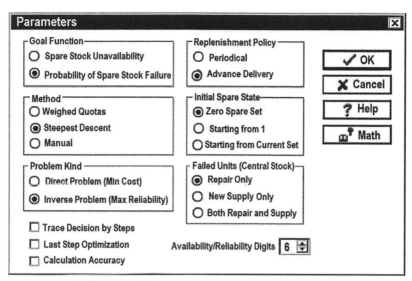

FIGURE 12.4 OSA tool: calculation options.

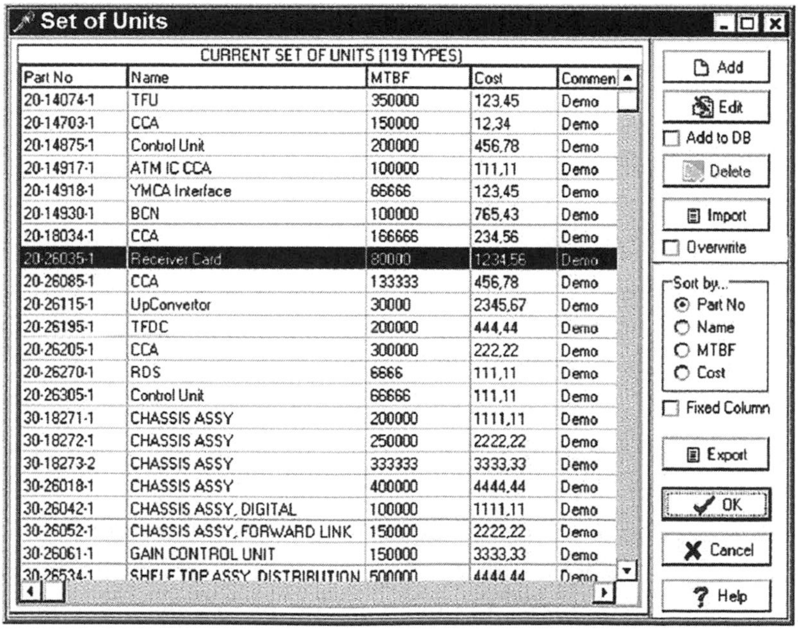

FIGURE 12.5 OSA tool: Unit database.

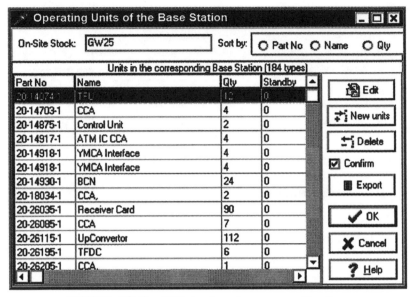

FIGURE 12.6   OSA tool: gateway specification.

how many active redundant units are installed. (The number of active redundant units influences the number of spare standby units.)

The report of the results is presented in a customized table form (Fig. 12.7). A report can reflect results for any particular OSS or any chosen set of them. (It gives an opportunity to choose the shape and size of regions for each RSS.)

The OSA tool is flexible and offers various calculation options to the user. It is able to solve the direct and inverse problems of optimal redundancy with two different goal functions. It also offers two separate replenishment policies. Results of calculations are presented in a report with layout specified by the user. Reports generated by the OSA tool may be saved for further processing or documentation.

OPTIMAL SPARE ALLOCATION
STOCKS

Stock: GW47	*Level:* On-Site				
Spare unit delivery time: 240				*Reliability:* 0,999988	
Rep air time:					
Unit data:					

Part No	Name	MTBF	Cost	Spare	Spare Cost
20-14074-1	TFU	350000	123,45	2	246,90
20-14703-1	CCA	150000	12,34	3	37,02
20-14875-1	Control Unit	200000	456,78	2	913,56
20-14917-1	ATM IC CCA	100000	111,11	3	333,33

FIGURE 12.7   OSA tool: sample of a report.

12.2   OPTIMAL CAPACITY DISTRIBUTION OF TELECOMMUNICATION BACKBONE NETWORK RESOURCES

This project was performed for American telecommunications company MCI Communications Corp[1] and was presented in 1991 at the MCI headquarters (Richardson, Texas). The objective of the project was to develop an algorithm that allowed optimal relocation of inner resources in case of failure of one of the main components of the MCI backbone network. The structure of the network and real data are given in conditional units.

12.2.1   Brief Description of the System

Consider an approximate and simplified backbone network. A backbone network is designed with some redundant capacity, so the network can stand periods of overloading and possible catastrophic failure due to natural cataclysms or intentional enemy action. To protect the entire system from lockout, each link of the

---

[1]MCI Communications Corp. was an American telecommunications company founded in 1963. In the beginning of 2006 MCI was incorporated into Verizon with the name Verizon Business.

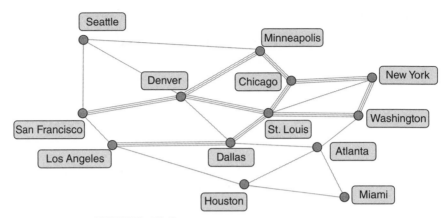

FIGURE 12.8   Backbone network structure.

network has a redundant capacity. Thus, in case of failures, neighbor links can take some traffic without jeopardizing their own successful operation. For instance, in Figure 12.8 triple lines denote links with higher capacity.

12.2.2   Conditional Example

A network structure itself is a highly redundant system. Consider, for instance, the connection San Francisco–New York (see Fig. 12.8). The direct path serves not only the chosen terminal nodes but also serves to interconnect all cities lying on this path: San Francisco, Denver, St. Louis, Chicago, and New York all connect each other.

All links have such capacity that normal connection can be achieved even in hours of maximum traffic. Most of the time only a relatively small portion of the links' capacity is in use, so the network has a significant redundancy.

Assume that the traffic matrix is known (in conditional units of traffic intensity; see Table 12.1). From this matrix, one can calculate the traffic at each link. In the result, we will have the traffic distri-

TABLE 12.1   Conditional Traffic Matrix

	S.F.	Denver	St. Louis	Chicago	N.Y.
S.F.	–	1	1	3	5
Denver	1	–	1	1	2
St. Louis	1	1	–	1	1
Chicago	3	1	1	–	8
N.Y.	5	2	1	8	–

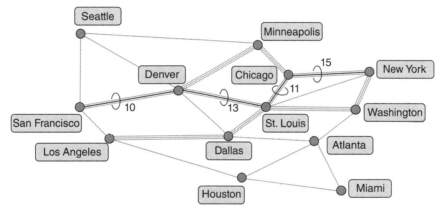

FIGURE 12.9   Path between San Francisco and New York with loading on the links.

bution in Figure 12.9 in the network operating state with no failures.

Let us consider the following scenario: the Denver–St. Louis link has broken down (see Fig. 12.10). What should be done to keep the normal connection between San Francisco and New York?

It is clear that the network having redundant capacity can redirect traffic through other links. However, a question arises: what is the best redirection of traffic?

To answer this question, one should formulate some requirements to the solution:

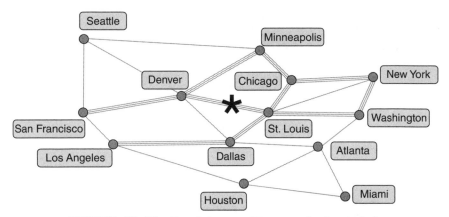

FIGURE 12.10 Breakdown of Denver–St. Louis link.

(a) The traffic from San Francisco to New York (and vise versa) should use the shortest available paths.

(b) A new path has to use high capacity links as much as possible.

(c) The traffic has to be redistributed in such a way that all used links have equal loading; that is, all links should have equal redundant capacity after rescheduling the traffic.

For finding the best redirection of the traffic, a linear programming model with some specific restrictions has been designed. The final solution of the traffic redirection is presented in Figure 12.11.

It is clear that a single link failure leads to traffic change in a significant part of the entire network. For instance, in the considered case we have to take into account not only traffic San Francisco–New York, but also traffic between other nodes laying on the original path. Actually, the model takes into account simultaneous traffic between all nodes of the network.

This model was intended to be used as a core controlling program for traffic redirection not only in cases of large break-

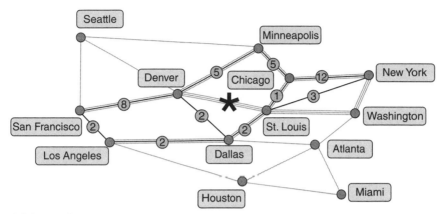

FIGURE 12.11   Example of redirection of traffic in the case of breakdown of the Denver–St. Louis link.

downs but also for temporary overloading of some parts of the network.

12.3   OPTIMAL SPARE ALLOCATION FOR MOBILE REPAIR STATION

This project was performed by contract with Hughes Network Systems (Germantown, Maryland) for a maintenance service of clients of a global telecommunication system. Names of units are changed and real data are rounded, though are kept in original range.

12.3.1   Brief Description of the System

There is a service base (SB) that serves terrestrially dispersed clients within a particular zone (see Fig. 12.12). All clients have almost similar equipment differing by configuration and scale. When equipment fails, a corresponding client sends a request for repair to the SB. Immediately after getting a request, SB directs an available mobile repair station (MRS) to the client.

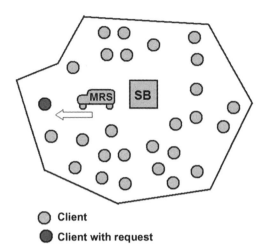

O Client
● Client with request

FIGURE 12.12 Schematic image of serving zone.

Each MRS has all needed instruments for repair and a set of spare field replacement units (FRU) for replacement of failed units. If there is a needed FRU, MRS performs a fast fix that takes just several minutes. Otherwise, a special request is sent to the SB and the needed FRU is delivered to the client in several hours. (In addition, it involves extra spending of money for restoration of the client's equipment.)

Equipment of clients can differ by configuration though it consists of almost the same set of components, the number of which exceeds several hundreds so it is impossible to keep FRU of all types in the MRS spare stock. Thus, one meets the problem of finding an optimal list of spares at MRS stock that provides maximum probability of first fix under the given restriction on the total available room for spares.

12.3.2 Formulation of the Problem

Denote available space of MRS stock V^*. Let client j, $j = \overline{1, M}$, have equipment with n_k^j components of type k (let's call it component k).

Denote failure rate of a component of type k by λ_k, $k = \overline{1, N}$. Then the flow of requests formed by components k, Λ_k, arriving at the SB, which can be written as

$$\Lambda_k = \lambda_k \sum_{1 \leq j \leq M} n_k^j. \tag{12.3}$$

The total flow of requests, Λ, is equal to

$$\Lambda = \sum_{1 \leq j \leq N} \Lambda_k. \tag{12.4}$$

It is clear that a current failure occurs due to a failure of component-k occurs with the probability

$$p_k = \frac{\Lambda_k}{\Lambda}. \tag{12.5}$$

Denote available space of MRS stock by V and physical volume of component k by v_k. If one assumes that there are no multiple instantaneous failures and the probability that the second failure of the same equipment during FRS travel time is negligibly small, than the solution of the problem is very simple: one calculates values

$$w_k = \frac{p_k}{v_k}, \tag{12.6}$$

and then takes the first S components that satisfy the following condition:

$$\sum_{1 \leq k \leq S} v_k \leq V < \sum_{1 \leq k \leq Sb1} v_k. \tag{12.7}$$

In practice, FRU of different types are approximately of the same volume, that is, $v_k = v$. It means that instead of ordering values w_k, it is enough to order values p_k. Refer to Figure 12.13.

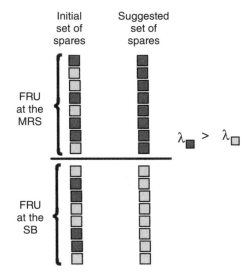

FIGURE 12.13 Explanation of the solution.

12.3.3 Case Study

In this particular case, the volumes of FRU are approximately the same, so the limitation is for the total number of FRUs, which is equal to 51.

Values of Λ_k for various components are given in Table 12.2. In this table, column "O" (for "old") contains the number of spares in the initial list and column "N" (for "new") contains the number of spares in the final list obtained by suggested method. For the sake of brevity, we omitted those types of equipment components for which both spare lists (initial and suggested) have zero spare units at the MRS stock.

From the complete list of equipment components (it is not presented), one can find that the total failure rate in the chosen service zone is equal to $\Lambda = 0.254$ (1/day), that is, approximately 1 failure in every 4 days. Failures covered by the initial set of spares form a failure flow with rate $\Lambda_k = 0.136$, and for a suggested set of spares the analogous value is equal to $\Lambda_k^{opt} = 0.229$.

TABLE 12.2   List of Redundant Units for Two Methods of Calculation

#	Part No.	Rate (per day)	O	N
1	30514	0.0149	0	1
2	15338	0.0118	1	1
3	14364	0.0114	0	1
4	17669	0.0107	1	1
5	30661	0.0103	2	1
6	19848	0.0096	1	1
7	11836	0.0095	1	1
8	13847	0.0093	0	1
9	35149	0.0088	1	1
10	12076	0.0086	0	1
11	17512	0.0077	1	1
12	30038	0.0071	1	1
13	16174	0.0069	1	1
14	11837	0.0057	0	1
15	30291	0.0056	1	1
16	17960	0.0053	0	1
17	92486	0.0053	0	1
18	13848	0.0052	0	1
19	17960	0.0050	0	1
20	19847	0.0045	1	1
21	37272	0.0042	0	1
22	15901	0.0041	0	1
23	18362	0.0041	1	1
24	19391	0.0039	1	1
25	30290	0.0038	0	1
26	14364	0.0035	1	1
27	30812	0.0032	0	1
28	15902	0.0031	0	1

(Continued)

TABLE 12.2 (*Continued*)

#	Part No.	Rate (per day)	O	N
29	12171	0.0025	1	1
30	11835	0.0023	1	1
31	19552	0.0023	0	1
32	17668	0.0023	0	1
33	10111	0.0022	0	1
34	92132	0.0022	1	1
35	70275	0.0021	0	1
36	12436	0.0020	1	1
37	11022	0.0018	1	1
38	12487	0.0018	1	1
39	11306	0.0017	0	1
40	11011	0.0016	1	1
41	20026	0.0015	1	1
42	30330	0.0015	1	1
43	30467	0.0014	1	1
44	92428	0.0012	1	1
45	90096	8.6E-4	1	1
46	30279	8.4 E-4	1	1
47	30140	6.8 E-4	1	1
48	11199	5.9 E-4	1	1
49	20079	4.9 E-4	1	1
50	30438	3.9 E-4	1	1
51	35111	3.6 E-4	1	1
52	15233	2.9 E-4	2	0
53	15187	2.9 E-4	1	0
54	30491	2.9 E-4	1	0
55	12256	2.4 E-4	1	0
56	17114	2.3 E-4	1	0
57	30510	2.2 E-4	1	0
58	93634	2.2 E-4	1	0
59	10306	1.9 E-4	1	0

TABLE 12.2 (Continued)

#	Part No.	Rate (per day)	O	N
60	17132	1.9 E-4	1	0
61	30470	1.9 E-4	1	0
62	30066	1.4 E-4	2	0
63	30206	1.4 E-4	1	0
64	30626	1.2 E-4	1	0
65	92513	1.2 E-4	1	0
66	11667	8.6E-5	1	0
67	20228	6.7E-5	1	0
68	11485	4.2E-5	1	0
69	35127	1.8E-5	1	0
70	11836	1.2E-5	1	0

It means that the probability of FF has been increased from

$$p_k = \frac{\Lambda_k}{\Lambda} = \frac{0.136}{0.254} \approx 0.535 \text{ to } p_k^{opt} = \frac{\Lambda_k^{opt}}{\Lambda} = \frac{0.229}{0.254} \approx 0.902.$$

12.3.4 Evaluation of Bargain

It interesting to evaluate what a financial bargain has to do with the application of simplest optimization technique. Approximate evaluation of expected gain made by customers showed that the entire service system spreading over the United States gets on average about 44,000 calls a year. MRSs with initial spare stocks made about 44,000 × (1 − 0.535) ≈ 20,500 extra deliveries due to lack of needed FRU in their stocks. The suggested spare stock leads only to 44,000 × (1 − 0.902) ≈ 4,300 extra deliveries, that is, about 16,500 deliveries fewer. Each visit takes on the average about 2 hours (round trip) and about 0.5 hour for equipment inspection at the client site. Each visit costs at least $150 (plus overheads), so the total gain is about $24.7 million a year.

CHRONOLOGICAL BIBLIOGRAPHY

Ushakov, I. A., and Aliguliev, E. A. 1989. "Optimization of data-transmitting network parameters using the method of statistical modeling." *Soviet Journal of Computer and System Sciences* (US), vol. 27, no. 1.

Ushakov, I. 1994. *Methods of Research in Telecommunications Reliability (An Overview of Research in the Former Soviet Union)*. RTA.

Antonov, S., Chakravarty, S., Hamid, A., Keliinoi, T., and Ushakov, I. 1999. "Spare supply system for Globalstar, a worldwide telecommunication system." *Proceedings of the 24th International Conference on Computers and Industrial Engineering*, Middlesex, England.

Chakravarty, S., and Ushakov, I. 1999. "Reliability Measure Based on Average Loss of Capacity." *Proc. of the 15th Triennial IFORS Conference*, Beijing, China.

Ushakov, I., Antonov, S., Chakravarty S., Hamid A., and Keliinoi, T. 1999. "Spare supply system for Globalstar, a worldwide telecommunication system." *Computers and Industrial Engineering*, no. 1.

Chakravarty, S., and Ushakov, I. 2002. "Effectiveness analysis of Globalstar gateways." *Proc. of the 2nd Biennial Conference Mathematical Models in Reliability*, Bordeaux, France.

Chakravarty, S., and Ushakov, I. 2002. "Reliability influence on communication network capability." *Methods of Quality Management*, no. 7.

Chakravarty, S., and Ushakov, I. 2002. "Reliability measure based on average loss of capacity." *International Transaction in Operational Research*, no. 9.

Puscher, W., and Ushakov, I. 2002. "Territorially dispersed system of technical maintenance." *Methods of Quality Management*, no. 2.

Puscher, W., and Ushakov, I. 2002. "Calculation of nomenclature of spare parts for mobile repair station." *Methods of Quality Management*, no. 4.

Ushakov, I.A. 2005. "Terrestrial maintenance system for geographically distributed clients." *The International Symposium on Stochastic Models in Reliability, Safety, Security and Logistics* (book of abstracts), Beer-Sheva.

CHAPTER 13

COUNTER-TERRORISM: PROTECTION RESOURCES ALLOCATION

This section is based on a presentation made by the author at the joint meeting of CREATE[1] and GAZPROM[2] delegation in Los Angeles in 2007. The meeting was dedicated to problems of defense against terrorists' attacks.

13.1 INTRODUCTION

For decades, the United States has focused its military and intelligence capabilities on potential enemies beyond its own borders. After September 11, 2001, it has become increasingly clear that our enemies have the ability and determination to reach through our defenses and strike at critical assets here at home.

[1] CREATE: National Center for Risk and Economic Analysis of Terrorism Events (at the University of Southern California, Los Angeles).
[2] GAZPROM is the largest Russian company and the world's biggest owner of oil and natural gas.

Optimal Resource Allocation: With Practical Statistical Applications and Theory,
First Edition. Igor A. Ushakov.
© 2013 John Wiley & Sons, Inc. Published 2013 by John Wiley & Sons, Inc.

Modern terrorism has gone from the frame of simple intimidation to the active destruction of the chosen country: the goal of terrorists' attacks is to cause the maximum possible material damage and/or human casualties. It is obvious that terrorists' activity is getting more and more organized and modern counter-terrorism is a real war with an invisible enemy.

The problem of protection of human beings, material objects, and political/historical objects of possible terrorist attacks arose. A defender usually spends more resources than terrorists, so optimal allocation of these resources is very important. It is clear that terrorists have many advantages: they choose the time of the attack, they choose the object of the attack, and they choose the weapon of destruction. In general, a defender does not know what terrorists' intentions are.

The proposed mathematical model was developed for optimal allocation of defenders' resources for best protection of the defended objects.

13.2 WRITTEN DESCRIPTION OF THE PROBLEM

13.2.1 Types of Counter-Terrorism Actions

What kind of protective actions against terrorists should we consider? They are in general as follows:

1. Safety: complex series of preventive measures creating a "counter-terrorist environment" in the country.
2. Survivability: complex series of measures developing special procedures to minimize human loss if the strike happens.
3. Pre-emptive measures for destroying the terrorist's ability to attack.

Safety includes a set of measures attempting to prevent terrorist actions (check points at airports, checking cargo, profiled visa control, registration of foreign visitors and control of their staying in the country, control over the purchase of dangerous components for composing bombs, etc.). The objective of these measures is to prevent the possibility of organizing the terrorists' acts by limiting the admittance of suspicious people in the country and by eliminating the possibility of collection/creation of a weapon of mass destruction (WMD).

Examples

(1) A soft visa control in the United States permitted a number of 9/11 terrorists to enter the country and to stay within it easily.

(2) Absence of document control permitted a group of foreign terrorists to get training in jet piloting that led to terrorists hijacking civil planes and directing them to the twin towers in New York and the Pentagon in Washington.

(3) Lack of control for purchasing of suspicious materials gave Americans McVeigh and Nichols in 1995 the chance to make an extremely destructive bomb and blast a governmental building in Oklahoma City, killing 168 innocent people and wounding about eight hundred (including children in a kindergarten).

Survivability includes a set of measures that help to lose fewer lives and to prevent public panic.

Example

In October of 2002, when Chechen terrorists held hostages at the Moscow theater, Russian counter-terrorist forces used poisonous gas against them. However, they were not supplied with the antidote, which led to an enormous loss of human lives.

Pre-emptive measures include political steps and economical steps.

Examples

(1) UN inspections of countries with possible cradling of terrorists.

(2) Embargo for states supporting terrorism.

(3) Direct military attacks on terrorists' bases as has been done against al-Qaeda in Afghanistan.

Of course, some unjustified actions (such as Bush's war in Iraq) could even increase terrorists' activity.

Our belief is that all these sides of the terror-fighting problem must be combined in an aggregated model, which can be used by decision makers of various positions.

Here, at the first step of modeling of counter-terrorism resources allocation, we will focus on the measure of protection of a single object, that is, on the safety problem. For this problem, one can formulate the following problems:

Direct Problem: Optimally allocate resources that guarantee a desirable level of protection of defended objects against terrorists' attacks with minimum possible expenses.

Inverse Problem: Optimally allocate available limited resources that guarantee the maximum possible level of protection of defended objects against terrorists' attacks.

Thus, there are two objective functions:

- Guarantee level of the object protection, and
- Cost of protective measures.

Different objects have different priorities (or values). For instance, a terrorist attack on a stadium during a performance or game might lead to huge loss of human life; an attack on a large bridge might create a serious communication problem for a relatively long time; the destruction of a national symbol might be a reason for widespread panic and a strong hit to the country's prestige.

It is assumed that counter-terrorism experts are able to formulate measures of priority, or "weights" of defended objects because without such priority, object defense is rather amorphous.

13.2.2 Definition and Notations

Let us assume that there are three distinct layers of objects safety protection: federal, state, and local (individual). All input data are assumed to be given by counter-terrorism experts. Let us introduce the following notations:

$F_i(\varphi_i)$: Subjective probability that an object within the country will be protected against terrorists attack of type i under the condition that on the federal level one spends φ_i resources. (Notice that this type of protection might be not applicable to all objects. For instance, increasing control of purchasing chemical materials for WMD design has no relation to possible hijacking.)

$S_i^{(k)}(\sigma_i^{(k)})$: Subjective probability that an object within state k will be protected against terrorist attack of type i under the condition that on the level of this particular state one spends $\sigma_i^{(k)}$ resources.

$L_i^{(k,j)}(\lambda_i^{(k,j)})$: Subjective probability that particular object j within state k (denoted as pair "k, j") will be protected against terrorists' attack of type i under the condition that one spends $\lambda_i^{(k,j)}$ resources.

$W^{(k,j)}$: "Weight" (or "measure of priority") of object (j, k).

13.3 EVALUATION OF EXPECTED LOSS

In this stage, we consider a single object, j, located in state k. Assume that only set $G_{k,j}$ of possible types of terrorists' attacks is possible against object (k, j). Under the condition of uncertainty, we have to assume that terrorists choose the most vulnerable object to strike. In this case, federal protection delivers to this particular object a level of safety equal to:

$$F^{(k,j)} = \min\{F_i, i \in G_{k,j}\}.$$

Now consider state k level. Using the same arguments, we can write for object (k, j) the level of protection delivered by the protective measures on the state level:

$$S^{(k,j)} = \min\{S_i, i \in G_{k,j}\}.$$

Assume that on a local level, object (k, j)'s protection is equal to $L^{(k,j)}$. (Postpone, for a while, how this value is obtained.) Then we can assume that measures of protection on all three layers (federal, state, and local) influence an object independently.

Let us, for the sake of concreteness, consider the safety of a stadium: federal measures are usually relatively rough and non-specific (general visa control, etc.), state measures are more specific (traffic control, attention to local communities' behavior, etc.), and local measures are focused on specific sides (police blocking of transportation, stronger patrolling, using dynamite-sniffing dogs, etc.). It is possible to say that the federal layer nets "large fish," the state layer can net "smaller fish," and, finally, the local layer nets even smaller though "very poisonous fish." So, the total probability of possible terrorist attack will be lessened by all three layers practically independently, that is, the probability of successful protection of object (k, j) can be found as:

$$P^{(k,j)} = 1 - (1 - F^{(k,j)}) \cdot (1 - S^{(k,j)}) \cdot (1 - L^{(k,j)}). \qquad (13.1)$$

Hence, the expected loss, $w^{(k,j)}$, from a possible attack in this case is equal to

$$w^{(k,j)} = W^{(k,j)}(1 - P^{(k,j)}). \qquad (13.2)$$

13.4 ALGORITHM OF RESOURCE ALLOCATION

Now we return to calculation of $L^{(k,j)}$ and to the problem of optimal allocation of resources for object (k, j) protection.

Consider $G_{k,j}$, a set of possible terrorists' actions against object (k, j). On the local layer we know functions $L_i(\lambda_i)$—subjective probability of protection of object (j, k) depending on spent resources λ_i for all possible types of terrorist attacks, where superscripts (k, j) are omitted, for the sake of simplicity. These functions are presented in Figure 13.1, where for illustration purposes only we depict only three such functions. (They should be defined by counter-terrorism experts.)

First, consider the direct problem: obtaining the desired level of safety due to measures on the local layer. If the chosen level is L^*, then each of functions $L_1(\lambda_1)$, $L_2(\lambda_2)$, and $L_3(\lambda_3)$ has to have its value not less than L^* because inequality

$$\min\{L_1(\lambda_1), L_2(\lambda_2), L_3(\lambda_3)\} \geq L^*$$

has to be held.

It is obvious that for the minimax criterion to have any $L_i(\lambda_i)$ larger than L^* makes no sense. So, the problem of protection

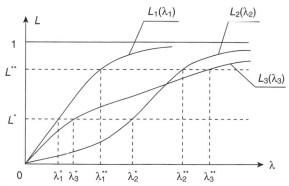

FIGURE 13.1   Examples of possible functions $L_i(\lambda_i)$.

resource allocation is solved: the local safety level L^* can be reached if all $L_i (\lambda_i) = L^*$, and in this case one spends total

$$\lambda^* = \lambda_1^* + \lambda_2^* + \lambda_3^*$$

resources. This amount of resources is the minimum for reaching safety level L^*.

In an analogous way, if one needs to reach the safety level L^{**}, the expenses related to this level of safety are

$$\lambda^{**} = \lambda_1^{**} + \lambda_2^{**} + \lambda_3^{**}$$

and also are the minimum for this case.

The inverse problem (maximization of safety under limited total resources) can be solved with the use an iterative process of numerical extrapolation. For instance, let total resources λ^0 be given. One can find two arbitrary solutions of the inverse problem, say, L^* and L^{**} with corresponding values λ^* and λ^{**}. Let all three values satisfy condition

$$\lambda^* \leq \lambda^\circ \leq \lambda^{**}.$$

Applying linear extrapolation, one finds value $L^{(1)}$ and then, having solved the Inverse Problem for this value, finds a new value $\lambda^{(1)}$, which is used on the second step of the iterative process instead of value λ^*, used at the beginning (see Fig. 13.2).

If initially found values λ^* and λ^{**} satisfy conditions $\lambda^* \leq \lambda^{**} \leq \lambda^0$ or $\lambda^0 \leq \lambda^* \leq \lambda^{**}$, obviously, the iterative process is absolutely similar.

Example 13.1

For the sake of transparency, consider a conditional example with stadium safety that gives us the possibility to explain everything

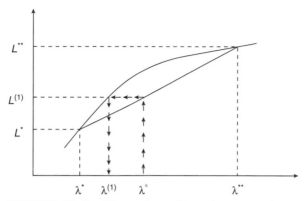

FIGURE 13.2 Illustration of iterative procedure.

less abstractly. Assume that protection measures on federal and state layers (for instance, attentive visa issuing with profiling nationality and country of applicant, checking pilot school attendees, observation of abnormal activity within specific communities, etc.) have been already undertaken.

Assume that three possible types of terrorist attacks are considered:

(A) Suicide bombing
(B) Track with explosive entering the stadium zone
(C) Crash of a private plane.

Assume that there are the following protection measures:

A_1: Visual checking of suspicious bags, dresses, and the like
A_2: Sample checking of suspicious persons
A_3: Using explosive-sniffing dogs
B_1: Police block of traffic on neighboring streets
C_1: Helicopter, armed with anti-plane missiles, barraging in the stadium area.

TABLE 13.1   Example Data

	Level of protection				
	0.9	0.95	0.99	0.995	0.999
A_1	1	2	5	8	12
A_2	5	10	25	40	60
A_3	–	2	–	5	10
B_1	1	–	10	–	20
C_1	50	75	125	200	300

Fictional numerical input data (expenses of these protective measures) used in this illustrative example is given in Table 13.1.

Here symbol "–" means that the protection level is absent; for instance, if one begins to use measure A_3, after applying 2 cost units the protection level jumps to 0.95, though there is no level 0.9 at all.

For the given example, expenses related to the protection level 0.95 are equal to 2+10+2+**10**+75=99 conditional units (numbers in the table are taken from most to least). Expenses related to level 0.995 are equal to 8+40+5+**20**+200=273 conditional units. Here bold fonts denote "jumps" described above, that is, one is forced to "overkill" protection since otherwise the required protection is not delivered.

We have outlined a very general theoretical approach that can be used for the assessing, planning, modeling, and managing of cost-effective counter-terrorism measures. The second phase of the proposed approach deals with an aggregated model for sets of defended objects within the states and in the country as whole. Of course, due to increasing dimensions of the problem on the higher layer it is possible to make only a computer model.

Having that computer model, one can formulate much more complex and realistic problems to include various "what if" scenarios and additional information: known gaps in security systems, counter-terrorism intelligence, impact of preemptive strike against terrorist groups, fuzzy (or not reliable enough) information about terrorist plans and capabilities.

Of course, among the protection measures, one should include secrecy of all undertaken defense activity, intentional propagation of disinformation about protective measures, and so on.

13.5 BRANCHING SYSTEM PROTECTION

Protecting the country against terrorist attacks cannot be accomplished without cost–performance analysis because of natural limitations on possible defending resources. The main problem for mathematical modeling of the phenomenon is its huge dimension.

Fortunately, the nature of the problem allows disaggregating the problem without loss of the sense of the problem. The system of a country defending target objects can be presented as a system with a special type of a branching structure with an additive type of global objective function.

The proposed approach assumes that input data are delivered by counter-terrorism experts.

13.5.1 Description of Levels of Protection

Counter-terrorism measures can be divided into three relatively independent levels in such a way that each level presents a kind of a sieve: the lower the level, the higher its "recognition," as shown in Figure 13.3.

On the federal and regional level the big "fish" are caught, and then after such sifting, chances of penetrating the three-level counter-terrorism protection barrier should be extremely small. Though errors in danger recognition, global protection actions, or

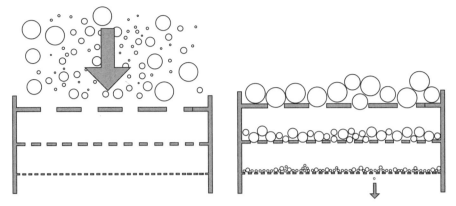

FIGURE 13.3   Explanation of different levels of protection.

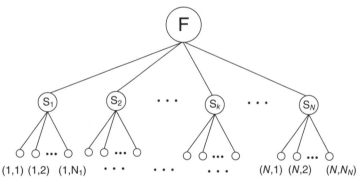

FIGURE 13.4   Three-level branching system.

preventive operations could lead to much more serious consequences than insufficient protection of local objects.

   Thus an adequate mathematical structure of such a protection
process is the so-called "branching structure" (see Fig. 13.4). The
upper layer is presented by federal counter-terrorism protection,
the middle layer consists of state-level protection, and defended
objects present the lower layer.

13.5.2 Model of Branching Structure

In this section we are using notation and terminology used in Section D. If for each object of the lower layer we choose an individual performance index, then the branching system performance index might be considered as a sum of individual indices. It follows from probability theory: mathematical expectation of the sum of random variables is equal to the sum of the mathematical expectations of random variables irrespective of dependence of the variables.

Indeed, introduce the so-called indicator function of the type:

$$\delta_{(k,j)} = \begin{cases} 1, & \text{if the attack on object } (k,j) \text{ has occurred,} \\ 0, & \text{otherwise.} \end{cases} \tag{13.3}$$

Then random loss for object (k, j) is equal to $\delta_{(k,j)} W^{(k,\,j)}$ and total random loss of all objects is

$$\sum_{k=1}^{N} \sum_{j=1}^{n_k} \delta_{(k,j)} W^{(k,j)}. \tag{13.4}$$

Mathematical expectation of this sum of random variables is defined as

$$w_{\text{Total}}\{F_i, \forall i; S_i^{(k)}, 1 \le k \le N; L_i^{(k,j)}, 1 \le j \le \hat{n}_k\} = E\left\{ \sum_{k=1}^{N} \sum_{j=1}^{Nk} \delta_{(k,j)} W^{(k,j)} \right\}$$

$$= \sum_{k=1}^{N} \sum_{j=1}^{Nk} E\{\delta_{(k,j)}\} W^{(k,j)} = \sum_{k=1}^{N} \sum_{j=1}^{Nk} (1 - P^{(k,j)}) W^{(k,j)} = \sum_{k=1}^{N} \sum_{j=1}^{Nk} w^{(k,j)},$$

$$\tag{13.5}$$

where $P^{(k,\,j)} = 1 - (1 - F^{(k,\,j)}) \cdot (1 - S^{(k,\,j)}) \cdot (1 - L^{(k,\,j)})$, and, in turn, these values are defined as

$$F^{(k,j)} = \min\{F_i, i \in G_{k,j}\}; \ S^{(k,j)} = \min\{S_i, i \in G_{k,j}\}; \ L^{(k,j)} = \min\{L_i\}.$$

In other words, Equation (13.5) gives the total expected loss with taking into respect the "weight" of each loss.

At the same time, it is easy to calculate the total expenses, C_{Total}, on all protection measures on all three layers:

$$C_{\text{Total}}\{\varphi_i, \forall i; \sigma_i^{(k)}, 1 \leq k \leq N; \lambda_i^{(k,j)}, 1 \leq j \leq n_k\}$$

$$= \sum_{\forall i} \varphi_i + \sum_{\forall i}\sum_{k=1}^{N} \sigma_i^{(k)} + \sum_{\forall i}\sum_{k=1}^{N}\sum_{i=1}^{n_k} \lambda_i^{(k,j)}. \tag{13.6}$$

Having objective Equations (13.5) and (13.6), one can formulate the following optimization problems:

Direct Problem
Optimally allocate total available resources C^0 that guarantee the minimum possible loss of defended objects against terrorists' attacks, that is,

$$\min\{w_{\text{Total}} \mid C_{\text{Total}} \leq C^0\} \tag{13.7}$$

Inverse Problem
Optimally allocate resources that guarantee the acceptable expected loss of defended objects against terrorist attacks with minimum possible expenses, that is,

$$\min\{C_{\text{Total}} \mid w_{\text{Total}} \leq w^0\}. \tag{13.8}$$

Solution of these problems with the use of the steepest descent method is demonstrated on a simple illustrative numerical example.

Example 13.2

Consider a fictional case study concerning embassy protection. There are three embassies within a geographical zone (Fig. 13.5). Embassies are assumed to have different indices according to prior-

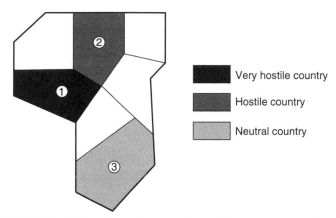

FIGURE 13.5 Allocation of embassies within a geopolitical region.

TABLE 13.2 **"Weight" of Importance = 10; Level of Protection with No Special Measures $P_1^{(0)} = 0.5$**

Safety	0.9	0.95	0.97	0.99
Expenses	2	4	7	12

TABLE 13.3 **"Weight" of Importance = 3; Level of Protection with No Special Measures $P_2^{(0)} = 0.8$**

Safety	0.9	0.95	0.97	0.99
Expenses	1	2	4	8

ity ("weights") and location in countries with different attitudes toward the country presented by the respective embassy. The problem is to protect all these embassies from terrorist attacks. Assume that there are available resources for embassy protection within each given zone (financial, military, logistics, etc.). How should they be allocated in the most reasonable way?

Let the following data be given by counter-terrorism experts. Assume that we consider only three embassies. The characteristics of these embassies are presented in Table 13.2, Table 13.3, and Table 13.4. The "weight" of importance might depend, for instance, on the size of the embassy (number of employees) or its political significance.

TABLE 13.4 "Weight" of Importance = 7; Level of Protection with No Special Measures $P_3^{(0)} = 0.9$

Safety	0.9	0.95	0.97	0.99
Expenses	0.5	1	2	5

Solution to Example 13.2

Calculate "discrete gradients" (relative increments) for each embassy k by using the formula:

$$\gamma_k^{(s)} = W_k \frac{P_k^{(s)} - P_k^{(s-1)}}{C_k^{(s)} - C_k^{(s-1)}}, \tag{13.9}$$

where W_k = "weight" of importance of the Embassy k, $P_k^{(s)}$ = level of protection at step s of the process of defense improving, and $C_k^{(s)}$ = expenses related to the level of protection at step s of the process of defense improving.

Let us construct Table 13.5, which will be used (in a very simple way!) to get an optimal allocation of money for defenses of all three embassies.

Now number all cells of Table 13.5 by decreasing values (Table 13.6). These numbers give the order of introducing corresponding protective measures. So, the final results are given as follows:

(1) Initial expected loss (no protection) is equal to

$$w^{(0)} = W_1 \cdot (1 - P_1^{(0)}) + W_2 \cdot (1 - P_1^{(0)}) + W_3 \cdot (1 - P_1^{(0)})$$
$$= \cdot 10 \cdot 0.5 + 3 \cdot 0.2 + 7 \cdot 0.1 = 3.8.$$

(2) After the first step, the total expected loss is equal to

$$w_{\text{Total}}^{(1)} = W_1 \cdot (1 - P_1^{(1)}) + W_2 \cdot (1 - P_1^{(0)}) + W_3 \cdot (1 - P_1^{(0)})$$
$$= \cdot 10 \cdot 0.1 + 3 \cdot 0.2 + 7 \cdot 0.1 = 1.8$$

and the spent resources are equal to $C^{(1)} = 2$.

TABLE 13.5   Values of Step-by-Step "Gradients"

No.	Step-by-step calculated value of "gradient"		
	Embassy 1	Embassy 2	Embassy 3
1	$10 \cdot \dfrac{0.9 - 0.5}{2} = 2$	$3 \cdot \dfrac{0.9 - 0.8}{1} = 0.3$	$7 \cdot \dfrac{0.95 - 0.9}{1} = 0.35$
2	$10 \cdot \dfrac{0.95 - 0.9}{4 - 2} = 0.25$	$3 \cdot \dfrac{0.95 - 0.9}{2 - 1} = 0.15$	$7 \cdot \dfrac{0.97 - 0.95}{2 - 1} = 0.14$
3	$10 \cdot \dfrac{0.97 - 0.95}{7 - 4} = 0.067$	$3 \cdot \dfrac{0.97 - 0.95}{4 - 2} = 0.03$	$7 \cdot \dfrac{0.99 - 0.97}{5 - 3} = 0.07$
4	$10 \cdot \dfrac{0.99 - 0.97}{12 - 7} = 0.004$	$3 \cdot \dfrac{0.99 - 0.97}{8 - 4} = 0.0015$	*

TABLE 13.6   Ordered Values of "Gradients"

No.	Value of "gradient" step-by-step		
	Embassy 1	Embassy 2	Embassy 3
1	1	3	2
2	4	5	6
3	8	9	7
4	10	11	*

(3) After the second step the total expected loss is equal to

$$w_{\text{Total}}^{(2)} = W_1 \cdot (1 - P_1^{(1)}) + W_2 \cdot (1 - P_1^{(1)}) + W_3 \cdot (1 - P_1^{(0)})$$
$$= \cdot 10 \cdot 0.1 + 3 \cdot 0.1 + 7 \cdot 0.1 = 1.5$$

and the spent resources are equal to $C^{(2)} = 2 + 1 = 3$.

(4) After the third step the total expected loss is equal to

$$w_{\text{Total}}^{(3)} = W_1 \cdot (1 - P_1^{(1)}) + W_2 \cdot (1 - P_1^{(1)}) + W_3 \cdot (1 - P_1^{(1)})$$
$$= \cdot 10 \cdot 0.1 + 3 \cdot 0.1 + 7 \cdot 0.05 = 1.15$$

and the spent resources are equal to $C^{(3)} = 2 + 1 + 1 = 4$.

(5) After the fourth step the total expected loss is equal to

$$w_{\text{Total}}^{(4)} = W_1 \cdot (1 - P_1^{(2)}) + W_2 \cdot (1 - P_1^{(1)}) + W_3 \cdot (1 - P_1^{(1)})$$
$$= \cdot 10 \cdot 0.05 + 3 \cdot 0.1 + 7 \cdot 0.05 = 0.9$$

and the spent resources are equal to $C^{(4)} = 2 + 1 + 1 + 2 = 6$.

(6) After the fifth step the total expected loss is equal to

$$w_{\text{Total}}^{(5)} = W_1 \cdot (1 - P_1^{(2)}) + W_2 \cdot (1 - P_1^{(2)}) + W_3 \cdot (1 - P_1^{(1)})$$
$$= \cdot 10 \cdot 0.05 + 3 \cdot 0.05 + 7 \cdot 0.05 = 0.75$$

and the spent resources are equal to $C^{(5)} = 2 + 1 + 1 + 2 + 1 = 7$.

(7) After the sixth step the total expected loss is equal to

$$w_{\text{Total}}^{(6)} = W_1 \cdot (1 - P_1^{(2)}) + W_2 \cdot (1 - P_1^{(2)}) + W_3 \cdot (1 - P_1^{(2)})$$
$$= \cdot 10 \cdot 0.05 + 3 \cdot 0.05 + 7 \cdot 0.03 = 0.61$$

and the spent resources are equal to $C^{(6)} = 2 + 1 + 1 + 2 + 1 + 1 = 8$.

(8) After the seventh step the total expected loss is equal to

$$w_{\text{Total}}^{(7)} = W_1 \cdot (1 - P_1^{(2)}) + W_2 \cdot (1 - P_1^{(2)}) + W_3 \cdot (1 - P_1^{(3)})$$
$$= \cdot 10 \cdot 0.05 + 3 \cdot 0.05 + 7 \cdot 0.01 = 0.47$$

and the spent resources are equal to $C^{(7)} = 2 + 1 + 1 + 2 + 1 + 1 + 2 = 10$.

(9) After the eighth step the total expected loss is equal to

$$w_{\text{Total}}^{(8)} = W_1 \cdot (1 - P_1^{(3)}) + W_2 \cdot (1 - P_1^{(2)}) + W_3 \cdot (1 - P_1^{(3)})$$
$$= \cdot 10 \cdot 0.03 + 3 \cdot 0.05 + 7 \cdot 0.01 = 0.37$$

and the spent resources are equal to $C^{(8)} = 2 + 1 + 1 + 2 + 1 + 1 + 2 + 3 = 13$.

(10) After the ninth step the total expected loss is equal to

$$w_{\text{Total}}^{(9)} = W_1 \cdot (1 - P_1^{(3)}) + W_2 \cdot (1 - P_1^{(3)}) + W_3 \cdot (1 - P_1^{(3)})$$
$$= \cdot 10 \cdot 0.03 + 3 \cdot 0.03 + 7 \cdot 0.01 = 0.31$$

and the spent resources are equal to $C^{(9)} = 2 + 1 + 1 + 2 + 1 + 1 + 2 + 3 + 2 = 15$.

(11) After the tenth step the total expected loss is equal to

$$w_{\text{Total}}^{(10)} = W_1 \cdot (1 - P_1^{(3)}) + W_2 \cdot (1 - P_1^{(3)}) + W_3 \cdot (1 - P_1^{(3)})$$
$$= \cdot 10 \cdot 0.01 + 3 \cdot 0.03 + 7 \cdot 0.01 = 0.21$$

and the spent resources are equal to $C^{(10)} = 2 + 1 + 1 + 2 + 1 + 1 + 2 + 3 + 2 + 5 = 20$.

The process of constructing tradeoff cost–protection can be continued. Graphical presentation of the steepest descent solution is presented in Figure 13.6.

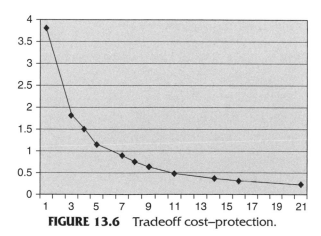

FIGURE 13.6 Tradeoff cost–protection.

This theoretical approach can be used for assessing, planning, modeling, and managing counter-terrorism measures. Some ideas of this approach were applied to modeling survivability of the National Energy System for the former USSR.

Further development of proposed theoretical approach and its implementation for various possible scenarios can significantly boost the analytic resources and predictive capabilities of fighting against terrorism. The approach is powerful enough for the solution of complex and highly unstructured problems. Based on this approach, one can formulate much more complex and realistic problems to include various "what if" scenarios and additional information: known gaps in security system, counter-terrorism intelligence, impact of preemptive strikes against terrorist groups, fuzzy information about terrorist plans and capabilities, and so on. Also, the proposed approach can be used to identify the most appropriate security measures and develop an optimal strategy aimed at providing maximum possible protection against terrorist threat. Finally, it may be useful in exploring the impact of budget cuts and resource reallocation scenarios on protection issues.

13.6   FICTIONAL CASE STUDY

13.6.1   Some Preliminary Comments

Before considering the fictional case study, let us once more briefly characterize the proposed model.

This mathematical model can be used for design of an interactive computer model, which can be used by counter-terrorism decision makers to solve the following problems:

- What are the priorities of the subjects of protection?
- What measures are most appropriate to protect these subjects?
- What is the best money allocation for this protection?

Notice from the very beginning that we are going to design a model for analysis of the entire problem as a unique and single "body."

A decision-maker will be able to "play" with the model, change parameters and limitations and get current results. In other words, it will be a "what if" type of model.

13.6.2 Suggested Procedure to Use the Model

An expert enters the following input data into the model:

- List of assumed objects of terrorists' attacks
- Priority of defended objects
- Estimated cost of various defending measures
- Estimated performance of the protection of listed subjects
- Limitation on the resources assigned for the protection program.

The model will present an output (solution) in the form of resource allocation between various measures of protection and between the chosen subjects of defense.

13.6.3 Input Data

- List of subjects of protection
- Categorization of the subjects of protection: human lives (at stadiums, conventions, etc.), economical, political, historical/symbolical
- Expert-prepared relative priorities of these subjects (on scale from 1 to 10, for instance)
- Assumed enemy's priorities of destruction of the same subjects
- Total resources for anti-terrorist activity on the country level and on regional levels

- Categorization of types of possible terrorist attacks for each subject of protection
- Expert's evaluation of the "degree of assurance" that the given subject of protection would be saved if some specified measures would have been undertaken
- Cost of reliable information about terrorists' plans, their location, forms of their support, and so on
- Experts' evaluation of the performance of the pre-emptive anti-terrorist strikes depending on the expenses.

Of course, the number of listed types of input information should be corrected (as well as the model itself) during real implementation.

Expected results of the modeling:

- The computer tool will allow a decision maker to estimate numerically the effect of counter-terrorism measures and will help to make optimal (rational) allocation of the available resources.
- The model will give the decision maker a range of human resources, finances, logistics, and so on needed for achieving the desired goal of protection of the chosen objects.

13.6.4   Description of the Subjects of Defense

Consider some fictional city, let's call it Freedom City, where there are the following subjects for counter-terrorism protection:

1. Stadium (during an event)
2. Monument of Glory
3. Great Bridge
4. Stock Exchange
5. National Park.

Let π_k be a priority number of subject k. The priority number in some sense reflects the priority of the subject for the society. Of course, such "scalar" evaluation is a trivialization of the problem, but this method is used in many cases by Operations Research analysts. The numerical value of π_k has to be decided by counter-terrorism experts. Let the priority numbers for the considering fictional example are:

$\pi_1 = 10$ (possibility of loss of huge number of lives)

$\pi_2 = 8$ (important national symbol)

$\pi_3 = 5$ (important transport link)

$\pi_4 = 7$ (destruction may lead to a large scale economic chaos)

$\pi_5 = 1$ (city's symbol, few living beings).

For the beginning, we do not pay attention to the specifics of the protected subject (human, economic, or political). These factors might be taken into account on the further stages of the research.

13.6.5 Description of types of possible attacks

Consider possible types of terrorists' attacks on the listed subjects and measures of protection.

Stadium

Possible types of attacks:

- Suicide bomber

   Form of protection:

   (1) Police at the entrance visually checking suspicious objects (big bags, suitcases, etc.); explosive-sniffing trained dogs are used.

   (2) Strong visa checking on the country boarders to avoid penetrating terrorists.

(3) Gathering information about unusual or suspicious activity within Freedom City community of possible origin of terrorist support.

(4) Emergency state for paramedic service of Freedom City.

• Private planes or helicopters used as kamikaze

Form of protection:

(1) Semi-military police helicopter barrage around the stadium with the weapons needed to destroy an unexpected flying object.

(2) Attentive scrutiny of candidates for pilot training schools.

(3) Emergency state for paramedic service of Freedom City.

• Regular civil planes

Form of protection:

(1) Hard checkpoints at airports (with inevitable politically incorrect profiling by name and appearance).

(2) Presence of marshals at each flight between large cities.

(3) Emergency state for paramedic service of Freedom City.

Monument of Glory

Possible types of attacks:

• Suicide bomber

Form of protection:

(1) Police at the entrance of the Monument of Glory to visually check suspicious objects (big bags, suitcases, etc.); using explosive trained dogs.

(2) Strong visa checking on the country boarders to avoid penetrating terrorists.

• Private planes or helicopters

Form of protection:

(1) Semi-military police helicopter barrage around the stadium with the weapons needed to destroy an unexpected flying object.

(2) Attentive scrutiny of candidates for pilot training schools.

· Regular civil planes

Form of protection:

(1) Hard checkpoints at airports (with inevitable politically incorrect profiling by name and appearance).

(2) Presence of marshals at each flight between large cities.

(3) Emergency state for paramedic service of Freedom City.

Great Bridge

Possible types of attacks:

· Suicide car bomber

Form of protection:

(1) Police at the entrance of the bridge checking suspicious vehicles.

(2) Police checking suspicious vehicles entering Freedom City.

(3) Strong checking of all employees at transportation organizations (with inevitable politically incorrect profiling by name and appearance).

(4) Strong visa checking on the country boarders to avoid penetrating terrorists.

· Bomb at the pier of the bridge

Form of protection:

(1) Control for suspicious diving activity in the area.

(2) Control for suspicious boat movement on the river under the Great Bridge.

(3) Strong checks of all owners of boats in the basin.

Stock Exchange

Possible types of attacks:

- Suicide car bomber

  Form of protection:

  (1) Police at the entrance of the street to the Stock Exchange checking suspicious vehicles.

  (2) Police checking suspicious vehicles entering Freedom City.

  (3) Strong checks of all employees at transportation organizations with inevitable profiling by appearance, name, and country of origin.

  (4) Strong visa checking on the country boarders to avoid penetrating suspected terrorists.

- Suicide bomber

  Form of protection:

  (1) Police at the Stock Exchange visually check suspicious objects (big bags, suitcases, etc.); using explosive-sniffing trained dogs.

  (2) Strong visa checking on the country boarders to avoid penetrating terrorists.

National Park

Possible types of attacks:

- Suicide bomber

  Form of protection:

  (1) Police at the entrance observe more carefully visitors to the National Park.

  (2) Strong visa checking on the country boarders to avoid penetrating terrorists.

From a brief analysis of measures above, one can see that some measures are local and specific for a particular subject of protection, some of them are common for objects within a particular region (or area), and, finally, some measures protect all subjects in the country.

For instance, the following two measures:

- police checking suspicious vehicles entering Freedom City and
- gathering information about unusual or suspicious activity within some communities

are local measures effective for the entirety Freedom City, but only for this city. At the same time, such measures as:

- strong visa checking at the country borders
- attentive attendee checking at the pilot training centers (with strong profiling of the trainees)
- closer checking at airports, especially to and from main cities, and
- putting marshals at the civil planes influence the subjects of protection within the entire country. All counter-terrorism pre-emptive strikes abroad always have an effect on the level of entire country.

13.7 MEASURES OF DEFENSE, THEIR EFFECTIVENESS, AND RELATED EXPENSES

In Table 13.7 we consider the simplest case study for the situation described above. All numbers are fictional. Expenses for various protection measures are taken in some conditional units (CU).

13.7.1 Federal Level of Protection

All expenses are to protect all objects within the country, so "individual" expenses for a single object will be relatively small. (For

TABLE 13.7   Measures of Protection at the Federal Level

Type of protection measure	Protection level		Related expenses	
Attentive visa issuing with respect to nationality and country of applicant	$F_{1(1)}$	0.9	C_{11}	5 CU
	$F_{1(2)}$	0.95	C_{12}	10 CU
Checking arrived passengers	$F_{2(1)}$	0.8	C_{21}	50 CU
	$F_{2(2)}$	0.9	C_{22}	100 CU
More attentive checking cargo	F_3	0.8	C_3	50 CU
Introducing strong control for staying in the country with guest visas	$F_{4(1)}$	0.8	C_{41}	300 CU
	$F_{4(2)}$	0.9	C_{42}	500 CU
Checking pilot school attendees	F_5	0.95	C_5	10 CU
Checking employees of transportation organizations	F_6	0.95	C_6	50 CU
. . . . .	. . . . .	. . . . .	. . . . .	. . . . .

instance, defending 10,000 objects within the country due to introducing strong control for staying in the country with guest visas will correspond approximately from 0.03 to 0.05 CU of expenses for each object.)

The list of possible types of protection measures on all levels, corresponding levels of protection, and related expenses have to be prepared by experts familiar with security problems.

13.7.2   Zone (Regional) Level of Protection

Every zone has to be considered individually because they have their own specific characteristics and their own measures of protection. Consider Freedom City as a zone and consider possible protection measures on the zone (regional) level (Table 13.8).

13.7.3   Local (Object) Level of Protection

All objects have to be considered individually because they have their own specific characteristics and their own measures of protec-

TABLE 13.8 Measures of Protection on the State Level

Type of protection measure	Protection level		Related expenses	
Checking incoming trucks	Z_1	0.9	C_{11}	5 CU
Police scrutiny of suspicious communities	$Z_{2(1)}$	0.8	$C_{2(1)}$	10 CU
	$Z_{2(2)}$	0.9	$C_{2(2)}$	15 CU
Control of airspace over the city	Z_3	0.95	C_3	50 CU
. . .	. . .	. . .	. . .	. . .

TABLE 13.9 Measures of Stadium Protection

Type of attack	Type of protection	Protection level		Related expenses	
Suicide bomber	Visual checking of suspicious personal belongings (bags, suitcases, etc.)	$L_{1(1)}^{(1)}$	0.9	$C_{1(1)}^{(1)}$	1 CU
		$L_{1(2)}^{(1)}$	0.95	$C_{1(2)}^{(1)}$	2 CU
	Sample checking of suspicious persons	$L_{2(1)}^{(1)}$	0.9	$C_{2(1)}^{(1)}$	3 CU
		$L_{2(2)}^{(1)}$	0.95	$C_{2(2)}^{(1)}$	4 CU
	Explosive–sniffing trained dogs	$L_3^{(1)}$	0.97	$C_3^{(1)}$	5 CU
Private plane crash	(Zone level)	–	–	–	–
Airliner crash	(Country level)				
. . . . .	. . . . .	. . . . .	. . . . .	. . . . .	. . . . .

tion. Besides, the same measures of protection might have different effects on different objects (Table 13.9, Table 13.10, Table 13.11, Table 13.12, and Table 13.13).

13.7.4 Calculation of Protection Level for Subjects of Defense

Demonstration of calculation of protection level will be performed on a single object, since calculation by hand is too time-consuming,

TABLE 13.10   Measures of Monument of Glory Protection

Type of attack	Type of protection	Protection level		Related expenses	
Suicide bomber	Visual checking of suspicious personal belongings (bags, suitcases, etc.)	$L_{1(1)}^{(2)}$	0.9	$C_{1(1)}^{(2)}$	0.5 CU
		$L_{1(2)}^{(2)}$	0.95	$C_{1(1)}^{(2)}$	1 CU
	Sample checking of suspicious persons	$L_{2(1)}^{(2)}$	0.9	$C_{1(1)}^{(2)}$	1.5 CU
		$L_{2(2)}^{(2)}$	0.95	$C_{1(1)}^{(2)}$	2 CU
	Explosive–sniffing trained dogs	$L_3^{(2)}$	0.97	$C_{1(1)}^{(2)}$	3 CU
Private plane crash	(Zone level)	–	–	–	–
Airliner	(Country level)	–	–	–	–
. . . . .	. . . . .	. . . . .	. . . . .	. . . . .	. . . . .

TABLE 13.11   Measures of Great Bridge Protection

Type of attack	Type of protection	Protection level		Related expenses	
Suicide car bomber	Police at the entrance of the bridge checking suspicious vehicles	$L_1^{(3)}$	0.95	$C_1^{(3)}$	1 CU
	Police checking suspicious vehicles entering Freedom City (zone level)	–	–	–	–
	Checking employees of transportation organizations (country level)	–	–	–	–
Bomb at the pier of the bridge	Control for suspicious boat movement on the river under the Great Bridge	$L_2^{(3)}$	0.99	$C_2^{(3)}$	3 CU
	Regular checking all owners of boats in the basin	$L_3^{(3)}$	0.95	$C_3^{(3)}$	1 CU

TABLE 13.12　Measures of Stock Exchange Protection

Type of attack	Type of protection	Protection level		Related expenses	
Suicide car bomber	Police at the street where stock exchange locates check suspicious vehicles parked at the building	$L_1^{(4)}$	0.95	$C_1^{(4)}$	0.5 CU
	Police checking suspicious vehicles entering Freedom City (zone level)	–	–	–	–
	Checking employees of transportation organizations (country level)	–	–	–	–
Suicide bomber	Visual checking of suspicious personal belongings (bags, suitcases, etc.) entering the building	$L_{2(1)}^{(4)}$	0.9	$C_{2(1)}^{(4)}$	0.5 CU
		$L_{2(2)}^{(4)}$	0.95	$C_{2(1)}^{(4)}$	1 CU
	Sample checking of suspicious persons	$L_{3(1)}^{(4)}$	0.9	$C_{3(1)}^{(4)}$	0.5 CU
		$L_{3(2)}^{(4)}$	0.95	$C_{3(1)}^{(2)}$	2 CU
	Explosive–sniffing trained dogs	$L_4^{(2)}$	0.97	$C_4^{(2)}$	1 CU

TABLE 13.13　Measures of National Park Protection

Type of attack	Type of protection	Protection level		Related expenses	
Suicide bomber	Visual checking of suspicious persons	$L_1^{(5)}$	0.95	$C_1^{(5)}$	1 CU
		$L_1^{(5)}$	0.95	$C_1^{(5)}$	1 CU
Strong visa checking on country borders	(Country level)	–	–	–	–

and in addition, a huge number of numerical results could make explanation not transparent.

Initial Level of Stadium Protection. For the Stadium, the value of expected minimax losses can be found as follows:

$$LOSS_{stadium} = \pi_1(1 - F_{1(1)}) \cdot (1 - F_{2(1)}) \cdot (1 - F_5) \cdot (1 - Z_3) \cdot (1 - L_{1(1)}^{(1)}) \cdot (1 - L_{2(1)}^{(1)}),$$

$$(13.10)$$

which after substitution of numerical values gives the result $LOSS_{stadium} = 0.0001$.

Corresponding expenses are: on the country level = 55 CU, on the zone level = 10 CU, and on the object level = 4 CU. Notice once more that the largest expenses are on the country level though these expenses are "shared" by all objects.

Taking into account only local (individual) factors, Equation (13.10) can be rewritten as follows:

$$LOSS_{stadium} = 0.01 \cdot (1 - L_{1(1)}^{(1)}) \cdot (1 - L_{2(1)}^{(1)}), \tag{13.11}$$

that is, we keep only variables related to the local level. Then, analyzing all possible measures, we will get the results presented in Table 13.14. In more visual form, the results are presented in Figure 13.7.

All these values can be used as members of the dominating sequences for further analysis. Such kind of analysis allows one to choose a balanced and effective allocation of resources between all three levels and to assign measures for each defended object depending on the possible type of terrorist attack.

TABLE 13.14   Results of Solutions

Variant no.	Formula for probability of loss calculating	Expected loss	Formula	Expenses
1	$(1 - L_{1(1)}^{(1)}) \cdot (1 - L_{2(1)}^{(1)})$	0.01	$C_{1(1)}^{(1)} + C_{2(1)}^{(1)}$	4 CU
2	$(1 - L_{1(2)}^{(1)}) \cdot (1 - L_{2(1)}^{(1)})$	0.005	$C_{1(2)}^{(1)} + C_{2(1)}^{(1)}$	5 CU
3	$(1 - L_{1(1)}^{(1)}) \cdot (1 - L_{2(2)}^{(1)})$	0.003	$C_{1(1)}^{(1)} + C_{2(2)}^{(1)}$	6 CU
4	$(1 - L_{1(2)}^{(1)}) \cdot (1 - L_{2(2)}^{(1)})$	0.0015	$C_{1(2)}^{(1)} + C_{2(2)}^{(1)}$	8 CU

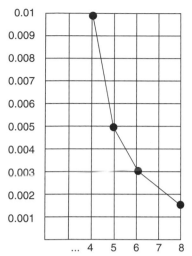

FIGURE 13.7   Tradeoff cost–protection.

13.8   ANTITERRORISM RESOURCE ALLOCATION UNDER FUZZY SUBJECTIVE ESTIMATES

13.8.1   Preliminary

The problem of optimal resource allocation for antiterrorism measures is naturally based on subjective estimates made by experts in this field. Relying on expert estimates is inevitable in this case: there is no other way to get input data for the system survivability analysis. There is no way to collect real data; moreover, there are no homogenous samples for consistent statistical analysis of observations, since any case is unique and nonreproducible. Nevertheless, quantitative analysis of necessary levels of protection has to be performed.

What are the subjects of such expertise? It seems to us that they are the following:

- Possibility of terrorist attacks on some object or group of objects
- Possible time of such attack

• Expected consequences of the attack and possible losses
• Possible measure of protection and related expenses.

Since expert estimates of such complex things are fuzzy due to lack of common understanding of the same actions and counter-actions within a group of experts, the question arises: is it possible at all to make any reasonable prognosis and, moreover, judgment about "optimal allocation of protection resources"?

First of all, we should underline that the concept of "optimal solution" relates only to mathematical models. In practice unreliable (and even inconsistent) data and inevitable inaccuracy of the model (i.e., the difference between a model and reality) allow us to speak only about "rational solutions."

Nevertheless, in practice the problem exists and in any particular case has to be solved with or without using mathematical models. Our objective is to analyze the stability of solutions of optimal resource allocation under the fuzziness of experts' estimates.

13.8.2 Analysis of Solution Stability: Variation of the Expense Estimates

First, let us analyze how variation of expenses estimates influences the solution of the problem on the level of a single object that has to be protested against terrorist attack. For transparency of explanation, we avoid considering the influence of defense on the federal and state levels.

Let us consider some conditional object (Object 1) that can be a subject of a terrorist attack. It is assumed that there may be three different types of enemy actions (Act. 1, Act. 2, and Act. 3). The defending side can choose several specific protection measures against each type of action {M(i, j), where i corresponds to the action type, and j corresponds to the type of undertaken protective measure.

Assume that we have three variants of estimates of protection measure costs: lower, middle, and upper as are presented in Table 13.15, Table 13.16, and Table 13.17. Here the lower estimates are about 20% lower of the corresponding middle estimates, and the upper ones are also about 20% higher.

There are three types of expert estimates: "optimistic," "moderate," and "pessimistic." The first ones assume that success in each situation can be reached by low expenses for protective measures, the last group requires larger expenses for protection in the same situation, and the middle group lies in between.

How can data in these tables be interpreted?

Let us consider the possibility of Act. 1 against the object. With no protection at all, the object's vulnerability equals 1 (or 100%). If one would have spent $\Delta E = 0.8$ conditional cost units (CCU) and undertook measure $M(1, 1)$, the object's vulnerability decreases to 0.25. If one does not satisfy such a level of protection, the next

TABLE 13.15   Case of Optimistic Estimates

OBJECT 1		γ_1	C
Act. 1	$M(1, 1)$	0.25	0.8
	$M(1, 2)$	0.2	2
	$M(1, 3)$	0.1	2.5
	$M(1, 4)$	0.01	3.8
Act. 2	$M(2, 1)$	0.2	1.6
	$M(2, 2)$	0.16	2.8
	$M(2, 3)$	0.07	3.2
	$M(2, 4)$	0.02	5.6
Act. 3	$M(3, 1)$	0.11	0.4
	$M(3, 2)$	0.1	2
	$M(3, 3)$	0.05	2.4
	$M(3, 4)$	0.04	3.6
	$M(3, 5)$	0.01	5.6

TABLE 13.16 Case of Moderate Estimates

OBJECT 1		γ_2	C
Act. 1	$M(1, 1)$	0.25	1
	$M(1, 2)$	0.2	2.5
	$M(1, 3)$	0.1	3
	$M(1, 4)$	0.01	4
Act. 2	$M(2, 1)$	0.2	2
	$M(2, 2)$	0.16	3
	$M(2, 3)$	0.07	4
	$M(2, 4)$	0.02	7
Act. 3	$M(3, 1)$	0.11	0.5
	$M(3, 2)$	0.1	2.5
	$M(3, 3)$	0.05	3
	$M(3, 4)$	0.04	5
	$M(3, 5)$	0.01	7

TABLE 13.17 Case of Pessimistic Estimates

OBJECT 1		γ_3	C
Act. 1	$M(1, 1)$	0.25	1.2
	$M(1, 2)$	0.2	3
	$M(1, 3)$	0.1	6
	$M(1, 4)$	0.01	7.8
Act. 2	$M(2, 1)$	0.2	2.4
	$M(2, 2)$	0.16	3.2
	$M(2, 3)$	0.07	4.8
	$M(2, 4)$	0.02	8.4
Act. 3	$M(3, 1)$	0.11	2
	$M(3, 2)$	0.1	3
	$M(3, 3)$	0.05	3.6
	$M(3, 4)$	0.04	6
	$M(3, 5)$	0.01	8.4

protective measure ($M(1, 2)$) is applied; that leads to decreasing the object vulnerability from 0.25 to 0.3 and costs 2 CCU.

Now let us consider all three possible terrorist actions. In advance nobody knows what kind of action will be undertaken against the defending object. In this situation the most reasonable strategy is providing equal defense levels against all considered types of terrorist attacks. It means that if one needs to ensure that a level of protection equals γ then one has to consider only such measures against each action that deliver a vulnerability level not less than γ. For instance, in the considered case, if the required level of vulnerability has to be no higher than 0.1, one has to use simultaneously the following measures of protection against possible terrorists' attacks: $M(1, 3)$, $M(2, 3)$, and $M(3, 2)$.

The method of equal protection against the various types of hostile attacks appears to be quite natural. If dealing with natural or other unintended impacts, one can speak about the subjective probabilities of impacts of some particular type of course; in a case of intentional attack from a reasonably thinking enemy, such an approach is not appropriate. The fact is that as soon as the enemy knows about your assumptions about his possible actions, he takes advantage of this knowledge and choose the hostile action that you expect least of all.

In the example considered earlier in this chapter, if one chooses measures $M(1, 2)$ with $\gamma_1 = 0.2$, $M(2, 3)$ with $\gamma_2 = 0.07$ and $M(3, 4)$ with $\gamma_3 = 0.04$, the guaranteed level of the object protection is

$$\gamma_{\text{Object}} = \max(\gamma_1, \gamma_2, \gamma_3) = \max(0.2, 0.07, 0.04) = 0.2.$$

For choosing a required (or needed) level of object protection, one can compile a function reflecting the dependence of vulnerability of protection cost (Table 13.18). This function is depicted in Figure 13.8. Without detailed explanations, we present numerical results for the cases of "moderate" (Table 13.19 and Fig. 13.9) and "pessimistic" (Table 13.20 and Fig. 13.10) estimates.

TABLE 13.18 Case of Optimistic Estimates

		Object 1	
Step number	Undertaken measures	Resulting γ_{Object}	Total Expenses, C_{Object}
1	$M(1, 1)$, $M(2, 1)$, $M(3, 1)$	max {0.25, 0.2, 0.11}=0.25	0.8+1.6+0.4=2.8
2	$M(1, 2)$, $M(2, 1)$, $M(3, 1)$	max {0.2, 0.2, 0.11}=0.2	2+1.6+0.4=4
3	$M(1, 3)$, $M(2, 2)$, $M(3, 1)$	max {0.1, 0.16, 0.11}=0.16	2.5+2.8+0.4=5.7
4	$M(1, 3)$, $M(2, 3)$, $M(3, 1)$	max {0.1, 0.07, 0.11}=0.11	2.5+3.2+0.4=6.1
5	$M(1, 3)$, $M(2, 3)$, $M(3, 2)$	max {0.1, 0.07, 0.1}=0.1	2.5+3.2+2=7.7
6	$M(1, 4)$, $M(2, 3)$, $M(3, 3)$	max {0.01, 0.07, 0.05}=0.07	3.8+3.2+2.4=9.4
7	$M(1, 4)$, $M(2, 4)$, $M(3, 3)$	max {0.01, 0.02, 0.05}=0.05	3.8+5.6+2.4=11.8
8	$M(1, 4)$, $M(2, 4)$, $M(3, 4)$	max {0.01, 0.02, 0.04}=0.04	3.8+5.6+3.6=13
9	$M(1, 4)$, $M(2, 4)$, $M(3, 5)$	max {0.01, 0.02, 0.01}=0.02	3.8+5.6+5.6=15

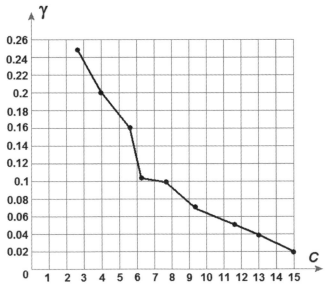

FIGURE 13.8 Dependence of the object survivability on cost of protection measures (for "optimistic" estimates).

TABLE 13.19 Case of Moderate Estimates

		Object 1	
Step number	Undertaken measures	Resulting γ_{Object}	Total expenses, C_{Object}
1	$M(1, 1)$, $M(2, 1)$, $M(3, 1)$	max {0.25, 0.2, 0.11}=0.25	1+2+0.5=3.5
2	$M(1, 2)$, $M(2, 1)$, $M(3, 1)$	max {0.2, 0.2, 0.11}=0.2	2.5+2+0.5=5
3	$M(1, 3)$, $M(2, 2)$, $M(3, 1)$	max {0.1, 0.16, 0.11}=0.16	3+3+0.5=6.5
4	$M(1, 3)$, $M(2, 3)$, $M(3, 1)$	max {0.1, 0.07, 0.11}=0.11	3+4+0.5=7.5
5	$M(1, 3)$, $M(2, 3)$, $M(3, 2)$	max {0.1, 0.07, 0.1}=0.1	3+4+2.5=9.5
6	$M(1, 4)$, $M(2, 3)$, $M(3, 3)$	max {0.01, 0.07, 0.05}=0.07	4+4+3=11
7	$M(1, 4)$, $M(2, 4)$, $M(3, 3)$	max {0.01, 0.02, 0.05}=0.05	4+7+3=14
8	$M(1, 4)$, $M(2, 4)$, $M(3, 4)$	max {0.01, 0.02, 0.04}=0.04	4+7+5=16
9	$M(1, 4)$, $M(2, 4)$, $M(3, 5)$	max {0.01, 0.02, 0.01}=0.02	4+7+7=18

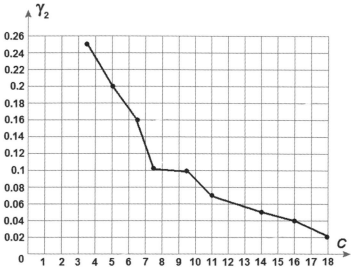

FIGURE 13.9 Dependence of the object survivability on cost of protection measures (for "moderate" estimates).

TABLE 13.20 Case of Pessimistic Estimates

	Object 1		
Step number	Undertaken measures	Resulting γ_{Object}	Total expenses, C_{Object}
1	$M(1, 1)$, $M(2, 1)$, $M(3, 1)$	max {0.25, 0.2, 0.11}=0.25	1.2+2.4+2=5.6
2	$M(1, 2)$, $M(2, 1)$, $M(3, 1)$	max {0.2, 0.2, 0.11}=0.2	3+2.4+2=7.4
3	$M(1, 3)$, $M(2, 2)$, $M(3, 1)$	max {0.1, 0.16, 0.11}=0.16	3+3.2+2=8.2
4	$M(1, 3)$, $M(2, 3)$, $M(3, 1)$	max {0.1, 0.07, 0.11}=0.11	3+4.8+2=9.8
5	$M(1, 3)$, $M(2, 3)$, $M(3, 2)$	max {0.1, 0.07, 0.1}=0.1	3+4.8+3=10.8
6	$M(1, 4)$, $M(2, 3)$, $M(3, 3)$	max {0.01, 0.07, 0.05}=0.07	4+4.8+3.6=12.4
7	$M(1, 4)$, $M(2, 4)$, $M(3, 3)$	max {0.01, 0.02, 0.05}=0.05	4+8.4+3.6=16
8	$M(1, 4)$, $M(2, 4)$, $M(3, 4)$	max {0.01, 0.02, 0.04}=0.04	4+8.4+6=18.4
9	$M(1, 4)$, $M(2, 4)$, $M(3, 5)$	max {0.01, 0.02, 0.01}=0.02	4+8.4+8.4=20.8

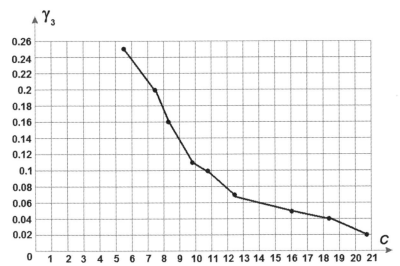

FIGURE 13.10 Dependence of the object survivability on cost of protection measures (for "pessimistic" estimates).

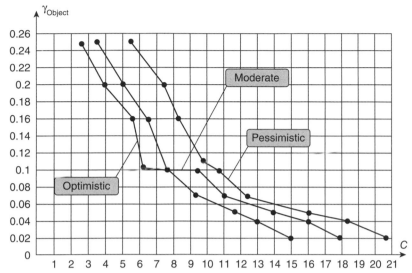

FIGURE 13.11 Comparison of solutions for three types of estimates.

Such analysis allows the possibility to find what measures should be undertaken for each required level of protection (or admissible level of vulnerability) and given limited resources. The final trajectory of the dependency of expenses versus vulnerability is presented in Figure 13.11. Using the the data presented for each estimate, consider two solutions of the direct problem with required levels of vulnerability 0.1 and 0.02.

However, decision makers are interested mostly in the correctness of undertaken measures, rather than in the difference in absolute values of the estimated costs of the object protection.

It is obvious that if the goal is to reach some given level of vulnerability the vector of the solution (i.e., set of undertaken measures for protecting the object against terrorist attacks) in the frame of considered conditions the vector of solution will be the same, though will lead to different expenses.

TABLE 13.21 Comparison of Solutions for Three Types of Scenarios

Type of scenario	Undertaken protection measures	
	$\gamma_{Object} \leq 0.1$	$\gamma_{Object} \leq 0.02$
Optimistic	$M(1, 3), M(2, 3), M(3, 2)$	$M(1, 4), M(2, 4), M(3, 5)$
Moderate	$M(1, 3), M(2, 3), M(3, 2)$	$M(1, 4), M(2, 4), M(3, 5)$
Perssimistic	$M(1, 3), M(2, 3), M(3, 2)$	$M(1, 4), M(2, 4), M(3, 5)$

Consider solutions for the required level of object vulnerability not higher than 0.1 and not lower than 0.02. They are presented in Table 13.21. One can see that solutions for all three scenarios coincide for both levels of object protection. Of course, such situations do not always occur, however, we should underline that vectors of the solution for minimax criterion $\gamma_{Object} = \max(\gamma_1, \gamma_2, \gamma_3)$ are much more stable than vectors for probabilistic criterion $1 - \gamma_{Object} = 1 - \prod_{1 \leq k \leq n}(1 - \gamma_k))$.

13.9 CONCLUSION

The presented analysis shows that the model of optimal allocation of counter-terrorism resources suggested in the previous section is working stably enough. Development of an improved computer model will allow analysis of more realistic situations, including random instability of input data. However, it seems that such "one-side biased" expert estimates can lead to more serious errors than random variations of the parameters.

CHRONOLOGICAL BIBLIOGRAPHY

Rudenko, Y., and Ushakov, I. 1979. "On evaluation of survivability of complex energy systems." *Journal of the Academy of Sciences of the USSR, Energy and Transportation*, no. 1

Kozlov, M., Malashenko, Y., Rogozhin, V., Ushakov, I., and Ushakova, T. 1986. "Computer model of energy systems survivability: methodology, model, implementation" (in Russian). *The Computer Center of the Academy of Sciences of the USSR, Moscow.*

I. Ushakov. 1994. "Vulnerability of Complex Systems." In *Handbook of Reliability Engineering.* ed. I. A. Ushakov. John Wiley & Sons.

Barbacci M. 1996. "Survivability in the age of vulnerable systems." *Computer,* no. 29.

Levitin, G., and Lisnianski, A. 2000. "Survivability maximization for vulnerable multi-state system with bridge topology." *Reliability Engineering and System Safety,* no. 70.

Levitin, G., and Lisnianski, A. 2001. "Optimal separation of elements in vulnerable multi-state systems." *Reliability Engineering and System Safety,* no. 73.

Levitin, G. 2003. "Optimal allocation of multi-state elements in linear consecutively-connected systems with vulnerable nodes." *European Journal of Operational Research,* no. 2.

Levitin, G. 2003. "Optimal multilevel protection in series-parallel systems." *Reliability Engineering and System Safety,* no. 81.

Levitin, G., and Lisnianski, A. 2003. "Optimizing survivability of vulnerable series-parallel multi-state systems." *Reliability Engineering and System Safety,* no. 79.

Levitin G., Dai Y., Xie M., Poh K. L. 2003. "Optimizing survivability of multi-state systems with multi-level protection by multi-processor genetic algorithm." *Reliability Engineering and System Safety,* no. 82.

Bier, V. M., Nagaraj, A. and Abhichandani, V. 2005. "Protection of simple series and parallel systems with components of different values," *Reliability Engineering and System Safety,* no. 87.

Korczak, E., Levitin, G. and Ben Haim, H. 2005. "Survivability of series-parallel systems with multilevel protection." *Reliability Engineering and System Safety,* no. 90.

Ushakov, I. A. 2005. "Cost-effective approach to counter-terrorism." *Communication in Dependability and Quality Management,* no. 3.

Ushakov, I., and Muslimov, A. 2005. "Cost-effective approach to counter-terrorism." *International Symposium on Stochastic Models in Reliability, Safety, Security and Logistics* (book of abstracts), Beer-Sheva.

Ushakov, I. A. 2006. "Counter-terrorism: protection resources allocation. Part I. Minimax criterion." *Reliability: Theory and Applications,* no. 2.

Ushakov, I. A. 2006. "Counter-terrorism: protection resources allocation. Part II. Branching system." *Reliability: Theory and Applications,* no. 3.

Bochkov, A. V., and Ushakov, I. A. 2007. "Sensitivity analysis of optimal counter-terrorism resources allocation under subjective expert estimates." *Reliability: Theory and Applications*, no. 2

Korczak, E., and Levitin, G. 2007. "Survivability of systems under multiple factor impact." *Reliability Engineering and System Safety*, no. 2.

Levitin, G. 2007. "Optimal defense strategy against intentional attacks." *IEEE Transactions on Reliability*, no. 1.

Ushakov, I. A. 2007. "Counter-terrorism: protection resources allocation. Part III. Fictional 'case study.'" *Reliability: Theory and Applications*, no. 1.

Hausken, K. 2008. "Strategic defense and attack for series and parallel reliability systems." *European Journal of Operational Research*, no. 2.

Hausken, K. 2008. "Strategic defense and attack of complex networks." *International Journal of Performability Engineering*, no. 4.

Hausken K., and Levitin, G. 2009. "Minmax defense strategy for complex multi-state systems." *Reliability Engineering and System Safety*, no. 94.

ABOUT THE AUTHOR

Professor Igor Ushakov graduated from Moscow Technical University (Moscow Aviation Institute), Electronics School, Department of

Optimal Resource Allocation: With Practical Statistical Applications and Theory,
First Edition. Igor A. Ushakov.
© 2013 John Wiley & Sons, Inc. Published 2013 by John Wiley & Sons, Inc.

Self-Controlled Surface-to-Air Missiles in 1958. In 1963 he earned his Ph.D. in Electronics, and in 1968 was awarded a Dr.Sc. (Doctor habilitatus) degree.

He has been working for over 40 years in industrial companies of the former Soviet Union and, later, in American companies as an employee and a consultant. Simultaneously, during that time he has headed departments at such prestigious Russian educational centers as Moscow Energy Institute and Moscow Institute of Physics and Technology (a Russian analogue to Massachusetts Institute of Technology) as well as held professorships (1989–2000) at American universities such as George Washington University, George Mason University, and the University of California at San Diego.

He is the author of about 30 monographs and over 300 scientific papers published in prestigious national and international scientific and engineering journals. He has also published eight books of prose, lyrics, and poems in Russian.

INDEX